Chemistry of Organometallic Compounds, Passivity and Corroision

Chemistry of Organometallic Compounds, Passivity and Corroision

Y.K. Chauhan

CYBER TECH PUBLICATIONS
4264/3, Ansari Road, Daryaganj, New Delhi-110002 (India)
Ph.: 011-23244078, 011-43559448 Fax: 011-23280028
E-Mail: cyberpublicationsdelhi@yahoo.com
Website: www.cybertechpublications.com

CHEMISTRY OF ORGANOMETALLIC COMPOUNDS, PASSIVITY AND CORROISION

Y.K. Chauhan

First Edition 2010

Published by : Asian Offset Printers

G.S. Rawat for **Cyber Tech Publications**
4264/3, Ansari Road, Daryaganj, New Delhi-110002 (India)
Ph.: 011-23244078, 011-43559448 Fax: 011-23280028
E-Mail: cyberpublicationsdelhi@yahoo.com
Website: www.cybertechpublications.com

PREFACE

This book is provides all necessary information and coverable solutions of Indian Universities. Solutions to the problem will be as diverse as the lands affected. However new, locally adapted and improved varieties will always be a central component in any package of engineering solutions. The subject matter has been presented in a simple and systematic manner. High and low temperatures, compounds with high levels of metal ions continue to reduce productivity over vast tracts of land and will remain an agricultural challenge for the foreseeable future.

Moreover in these times of surpluses in developed countries, solutions to the problem of abiotic stress are laid firmly at the door of developing country agriculture.

This book is devoted completely to the researches and their outcomes in the field of crop and plant physiology, undersigned is confident enough that this work of him would give readers knowledge abouth these researches.

—Author

PREFACE

Author

Contents

1

Classification of Organometallic Compounds

INTRODUCTION

Types of Metal-Carbon Bonds

The reactively and the structures of those compounds are best related to the character of the metal-carbon bond.

COVALENT ORGANOMETALLIC COMPOUNDS (σ-BONDED)

Generally the non-transition elements form the σ-bonded covalent organometallic compounds. The polarity of the bond depends not only on the difference in electro-negativity between metal atom and carbon but also on the metal orbitals which are available, the steric requirements and the polarisability of the electron cloud *e.g.*,

$$B(CH_3)_3;\ Al(CH_3)_3;Si(CH_3)_4;\ P(C_6H_5)_3.$$

Transition metals can also form metal-carbon σ-bonds with organic groups provided some additional π-bonding ligands are also present in the molecule, attached directly to the metal atom *e.g.*,

$$CH_3Mn(CO)_5.\pi.C_6H_6Fe(CO)_3R.$$

Binary alkyls of transition elements are more stable than binary alkyls of transition elements which decompose even at room temperature.

These compounds have predominantly 2-electron-2-centered σ-bond formed between the organic group and the atom M, though the bonds have much ionic character due to the electronegativity differences. These compounds are formed mostly by the nonmetallic and metalloid elements: Me_3SnX . $SiMe_2Cl_2$, $Pb(C_2H_5)_4$, $(CH_3)_2Zn$; etc.

Normal valence rules apply to the bonding and the structure of these compounds, and the alkyl group present can be substituted by other monodentage groups (halogen atoms, OH^- etc.). Thus, in the structure of $(CH_3)Si$ the four methyl groups are occupying the four corners of a tetra-hedron while silicon is situated at the centre. In these compounds, the alkyl bridges are present in which the carbon atom of the bridging alkyl group is equidistant from both the metal atoms.

The nature of bonding in these bridging alkyl group complexes cannot be explained on the basis of the conventional 2-electron-2 centre bonds, as carbon atom which is already attached to three hydrogen (or carbon) atoms, cannot be bonded to two M atoms simultaneously. The bonding is explained on the basis of the MOT in which M–C–M bond is formed by the use of sp^3 hybrid orbitals of carbon atoms and of each one of the M atoms.

ORGANOMETALLIC COMPOUNDS (IONIC)

The hydrocarbon is present as a carbon which being negatively charged is strongly attracted to the metal anion by non-directional electrostatic forces *e.g.,* sodium and heavier alkali metallic compounds. $C_{10}H_8Na$ ($C_{10}H_8$ group is naphthalene radical anion and is paramagnetic). Organo sodium compounds are generally colourless salt-like solids, which are electrolytes in fused form and are insoluble in non-polar solvents. Organometallic compounds show a gradual transition from ionic to σ^- bonded covalent forms.

Thus, alkyls of sodium are more ionic and less covalent than those of lithium. These are extremely reactiye and prepared in ethereal solvents: *e.g.,* diethyl ether, tetra-hydrofuran, diglyme. These compounds are non-isolable as solids.

These compounds are unstable in air, are readily hydrolyzed in water to liberate the hydrocarbon with the formation of hydroxide, and are insoluble in the organic hydrocarbon solvents. Their stabilities depend on the stability of the organic radical R^-. Thus, $C_6H_5Na^+$ and $C_5H_5Na^+$ are stable due to the stabilization of the negative charge through the π electron density.

The structures of ionic compound depend upon the manner of packing in the crystal; for example in the structure of $Mn(C_6H_5)_2$ each Mn_2^+ ion is situated midway between two parallel $C_6H_5^-$ ions.

CARBON METAL BOND NATURE

Metals

A metal is considered to be an element which is good conductor of heat and electricity and has metallic lustre. In chemical sense metal is an electro positive element which possesses tendency to lose electrons in the chemical combination and yields positive ions. All the elements could be divided into three classes on the basis of electro-negativity :

(a) Metals,

(b) Non-metals,

(c) Metollaids.

From the electro-negativity data, it is concluded that elements having electro-negativity less than 1.8 are considered as true metals and the elements having electro-negativity in the range 1.8-2.0 are considered as metalloids. In organometallic chemistry, metal is defined as an element which is much more electropositive than carbon, the electro - negativity of carbon being 2.5.

POLARITY OF CARBON METAL BONDS

As all the metals forming organo metallic compounds, possess electro-negativity less than carbon, they will be strongly positive with respect to carbon. Thus we will regard all organometallic substances as ionic with carbon always negative with respect to the metal.

In organometallic chemistry we are not concerned with carbon metal bonds as such but rather with alkyl, aryl, unsaturated hydrocarbon, heterocyclic and substituted organic groups attached to the various metals. The organic groups attached to the-metal become different in their effective electro-negativities, relatively more positive by the presence of much hydrogen or more negative by lacking of hydrogen. The organic group may be still more negative if fluorine, chlorine, oxygen or nitrogen are present.

Thus, it may be said that the ionic character of carbon metal bonds gets increased if the carbon is a part of an unsaturated or aromatic system and gets greatly increased when the, organic groups carry halogens, nitrogen or oxygen as substituents. Since fluorine is most electro-negative, the fluorocarbon compounds such as CF_3MgI are too polar as to have a little solubility in organic solvents.

The electro-negativity difference may be so great as to form salt like compounds having discrete positive and negative ions which are held together by ionic bonds *e.g.*, sodium cyclo-pentadienide $C_5H_5Na^+$ which gives rise to the reactions of an ionic substance and has a lattice structure.

Non Ionic Bonds to Transition Metals

There are some organic molecules which are sufficiently electron rich and may donate electron pairs resulting in the formation of coordination bonds. It is not possible to classify the actual bond formed as ionic or covalent or the mixture of two.

The electrons are frequently delocalized so that we can no longer think of individual metal carbon bonds. Such bonds formed between hydrocarbon groups and many of the transition metals are known as "sandwich bonds". From this it is expected that organometallic compounds involve ionic covalent and delocalised electron bonds and all possible mixture of the three carbon bonds.

Multiple Bonding

If there occurs multiple bonding between carbon and metals, the bond strength would be increased with a little change in its polarity, p–$R_3M–C_6H_4–COOH$ is an example of multiple bonding through the use of d-orbitals. In this π-electrons in the benzene ring have been contributed to the d-orbitals of metal forming a $d_\pi - p_\pi$ bond which increases the strength of the bond between metal and ring. It is evident from the contribution of electrons from carbon atom and its associated ring to vacant d-orbitals of silicon.

Electrons from tilled d-orbital of metal may also be contributed to partly filled p-orbitals of carbon atoms in carbon monoxide yielding a bond typical in metal carbonyls.

Methods of Preparation

The various methods used for the preparation of organometallic compounds can be classified into three categories :

(a) Formation by addition reactions.

(b) Formation by substitution reactions.

(c) Miscellaneous methods.

Let us discuss these one by one.

FORMATION BY ADDITION REACTIONS

(a) Methods involving free radicals : The general reaction may be put as follows :

$$R + M \rightarrow R - M$$

The free methyl radicals were produced by the pyrolysis of tetra methyl lead and were made to treat with metals like Fe, Sn, Zn to form the corresponding methyl compounds.

$$Pb(CH_3)_4 \rightarrow Pb + 4CH_3$$

$$Sn + 4CH_3 \rightarrow Sn(CH_3)_4$$

$$Zn + 2CH_3 \rightarrow Zn(CH_3)_2$$

$$Fe + 2CH_3 \rightarrow Fe(CH_3)_2$$

The same reaction occurs with stable radicals if the metal is sufficiently electropositive.

$$Ph_3C + Na \rightarrow Ph_3C^-Na^+$$

(b) Reaction of CO with metals : The metals such as Ni, Fe, Co etc., react under controlled conditions with CO to form carbonyls.

$$Fe + 5CO \xrightarrow[200\text{ atm. pressure}]{200^\circ C} Fe(CO)_5$$

$$2Co + 8CO \xrightarrow[100\text{ atm. pressure}]{150^\circ C} Co_2(CO)_8$$

$$Ni + 4CO \xrightarrow[\text{and pressure}]{\text{Ordinary pressure}} Ni(CO)_4$$

(c) Reaction of metals with olefins or substituted alkanes ;

Sodium and other alkali metals add to olefins thereby forming ionic organometallic compounds.

$$Ph_2C = CPh_2 + 2Na \rightarrow \underset{\displaystyle Na}{\underset{|}{Ph_2C}} - \underset{\displaystyle Na}{\underset{|}{CPh_2}}$$

With certain acetylenes, two atoms of sodium add to two molecules of hydrocarbon in the following way :

$$2PhC \equiv CPh + 2Na \rightarrow \underset{\displaystyle Na}{\underset{|}{PhC}} = CPh - CPh = \underset{\displaystyle Na}{\underset{|}{CPh}}$$

Substituted ethanes react with alkali metals when at least two aryl groups get attached to each carbon atom.

$$Ph_2CHCHPh_2 + 2K \rightarrow 2Ph_2CH^-K^+$$

$$Ph_3CPh_3 + 2Na(Hg) \rightarrow 2Ph_3C^-Na^+$$

(d) Hydrometallation or Addition of metal hydrides to unsaturated hydrocarbons : The general reaction is

An example is the hydroboration reactions :

$$3RCH = CH_2 + BH_3 \xrightarrow{\text{ether}} (RCH_2CH_2)_3B$$

Borane, BH_3, is generated in the presence of the alkene by treating sodium borohydride ($NaBNH_4$) with boron trifluoride (BF_3). If it is carried out in the absence of ether, then the reaction must be heated to 100°C and still takes several days to form the alkylboron product. However, in the presence of ether, The reaction is very efficient at room temperature. Since this discovery in the 1950s the chemistry of organoboranes has fluorished, thanks largely to the pioneering work of H.C Brown (who was awarded the Noble prize for Chemistry in 1979 for his work in this field). Compounds containing silicon-hydrogen bonds add to alkenes at high temperatures in the absence of a catalyst (or at more moderate temperatures with a periodic catalyst):

$$HSiCl_3 + C_6H_{13}CH = CH_2 \rightarrow C_6H_{13}CH_2CH_2SiCl_3$$

Ethyne can undergo two successive additions of this kind :

$$HSiCl_3 + HC \equiv CH \rightarrow CH_3 = CHSiCl_3$$

$$CH_2 = CHSiCl_3 + HSiCl_3 \rightarrow Cl_3Si - CH_2CH_2 - SiCl_3$$

Aluminium hydride (AlH_3) is another reactive hydride that will add to alkenes, but it is quite difficults to make and is so reactive that it is not easy to store or handle. The difficulties can be circumvented via a reaction developed by Ziegler in the 1950s and now used industrially to prepare alkylaluminium compounds on a large scale (10^5 tonnes per annum) for use as polymerisation catalysts. The overall reaction can be written as a direct synthesis in which metallic aluminum reacts with a mixture of hydrogen and with a mixture of hydrogen and alkene gases at high pressure, for example :

$$2Al + 3H_2 + 6C_2H_4 \rightarrow 2(C_2H_5)_3Al$$

The reaction has been shown to occur in several stages, and if it is stopped before completion, diethyl aluminium hydride is obtained. The reaction can therefore be expanded as :

$$2Al + 3H_2 + 4C_2H_4 \rightarrow 2(C_2H_5)_2AlH$$

$$(C_2H_5)2AlH + C_2H_4 \rightleftharpoons (C_2H_5)_3Al$$

The reversibility of the second reaction is important because it allows a series of trialkylaluminium compounds to be prepared from one starting trialkyl compound. 2-methylpropene reacts more readily than ethene and it is easier to displace, so this is often used to form the initial trialkyl compound (Scheme 1). Reactions of this kind are used to produce these industrially important alkyl compounds, which are all highly reactive liquids that are used extensively a catalysts in the polymerisation of alkenes.

$$\begin{array}{ll} & \quad CH_2 = CH_2 \\ & \qquad\quad | \qquad\qquad (C_2H_5)3Al \\ 2Al + 3H_2 + 6(CH_3)2C = CH_2 \rightarrow & 2(CH_3)2CH(CH_2)3Al \\ & \qquad\quad | \qquad\qquad (C_3H_7)Al \\ & \quad CH_3CH = CH_2 \end{array}$$

Scheme I

(v) Addition of metal alkyls to unsaturated hydrocarbons :

Metal alkyls add to olefins to form organometallic compounds.

$$Et_3Al + CH_3 = CH_2 \rightarrow Et_2AlCH_2 - CH_2t$$

$$Ph_2C = CH_2 + Bu^nLi \xrightarrow{\text{Diethyl}} Bu^nCH_2CPh_2^- \; Li^+$$

$$Et^2Be + Me\,Pr^n\,C + CH_2 \xrightarrow{\text{ether}} MeEt\,Pr^n\,C.CH_2BeEt$$

SUBSTITUTION METHODS

(a) *Reaction of Metals with hydrocarbons :* Hydrocarbons having one or more hydrogen atoms which are acidic or active enough to behave like hydrogen atoms undergo reaction with highly electropositive metals to form organometallic compounds

$$RC \equiv CH + Na \rightarrow RC \equiv CNa + 1/2H_2$$

Triphenyl methane is another hydrogen compound which reacts with sodium in the following way :

$$\begin{array}{ccc} C_6H_5 & & C_6H_5 \\ | & & | \\ 2C_6H_5 - CH + & 2Na \rightarrow 2C_6H_5 - & CH - Na + H_2 \uparrow \\ | & & | \\ C_6H_5 & & C_6H_5 \end{array}$$

Cyclopentadiene undergoes reaction with metals such as potassium and iron to form organometallic compounds.

$$Fe + 2C_5H_6 \rightarrow (C_5H_5)_2\ Fe + H_2 \uparrow$$

Cyelopentadiene is so reactive that it is able to specially prepared iron oxide or copper oxide.

$$FeO + 2C_5H_6 \xrightarrow[\text{Phosphine}]{\text{Triethyl}} (C_5H_5)_2\,Fe + H_2O$$

$$CuO + 2C_5H_6 \xrightarrow[\text{Phosphine}]{\text{Triethyl}} (C_5H_5)_2\,Cu + H_2O$$

(b) *Reaction of halocarbons with Metals :* Alkyl halides under go reaction with metals under suitable conditions to form organometallic compounds. Thus, zinc diethyl is obtained by heating the ethyl iodide with zinc coppcr couple and distilling in an carbon dioxide. The better yields are obtained if we carry out the distillation a vacuum.

$$C_2H_5I + Zn \rightarrow C_2H_5ZnI$$

$$2C_2H_5ZnI \rightarrow (C_2H_5)_2Zn + ZnI_2$$

Grignard used the method for preparing magnesium compounds as

$$CH_3I + Mg \xrightarrow[\text{ether}]{\text{dry}} CH_3MgI$$

Organolithium compounds are also prepared by similar methods. Their general reaction is

$$RX + 2Li \rightarrow RLi + LiX$$

The reaction could be carried out in light petroleum, cyclohexane, ether or benzene in an atmosphere of dry nitrogen. In the aliphatic series the order of reaction has been as follows

chlorides > bromides > iodides

With the exception of methyl iodide, iodides can not be used because they react readily with organolithium compounds formed.

$$RLi + R'I \rightarrow R - R' + LiI$$

Hence, p-bromodimethyl aniline yields $Me_2NC_6H_4Li$ in about 95% yield.

$$Me_2NC_6H_4Br \rightarrow Me_2NC_6H_4Li + LiBr$$

(c) *Reaction of metals on Ethers :* The general reaction may be put as follows :

$$ROR' + 2M \rightarrow R - M + MOR'$$

The method is especially used for the preparation of organometallic compounds which are so reactive and react further with alkyl halides. Hence benzyl and benzyl hydryl methyl ethers react readily to give good yields.

$$PhCH_2OMe + 2Li \xrightarrow{T.H.F.} PhCH_2Li + LiOMe$$

$$PhC.OMe_2OMe + Na - K \xrightarrow{Pentane} PhCMe_2Na + KOMe$$

(d) *Transmetallation or Metal* : Metal Exchange Reactions : Metal-metal exchange reaction provides the most convenient method for the preparation of various organometallic compounds.

The general reaction scheme for transmetallation is :

$$RM' + M \rightleftarrows RM' + M$$

where M and M' represent two different metals. The reaction could go in either direction and the preparative value depends on the position of equilibrium. In very general terms, a more reactive metal will displace a less reactive one. In practice the most versatile reagents Rm in such reactions are the derivatives of 'heavy' metals such as mercury (Hg), thallium (Tl), lead (Pb) or bismuth (Bi). The most widely used are the organomercury compounds R_2Hg. Metals that can be alkylated or arylated by organomercury compounds include lithium, sodium, magnesium, calcium, zinc aluminium, gallium, tin, antimony, selenium, and tellurium. Some typical reactions are :

$$(CH_3)_2Hg + 2Na \rightarrow 2CH_3Na + Hg$$

$$(CH_3)_2Hg + Zn \rightarrow (CH_3)_2Zn + Hg$$

$$3(CH_3)_2Hg + Al \rightarrow 2(CH_3)_3Al + 3Hg$$

$$2(CH_3)_2Hg + Sn \rightarrow (CH_3)_4Sn + 2Hg$$

The mercury formed in each of these reactions collects as a liquid at the bottom of the reaction vessel and further part in the reactions of the new organometallic reagent if the latter is used directly. Similarly, Grignard's reagent reacts with various compounds to form organometallic compounds.

$$HgCl_2 + 2C_2H_5MgBr \rightarrow (C_2H_5)_2Hg + MgCl_2 + MgBr_2$$

$$2PbBr_2 + 4C_2H_5MgBr \rightarrow (C_2H_5)_4Pb + 4MgBr_2 + Pb$$

$$CdCl_2 + 2CH_3MgBr \rightarrow (CH_3)_2Cd + MgCl_2 + MgBr_2$$

$$ZnBr_2 + 2C_2H_5MgBr \rightarrow (C_6H_5)_2Zn + MgBr_2$$

$$SiCl_4 + 4C_2H_5MgBr \rightarrow (C_6H_5)_4Si + 4MgCl_2$$

Metal cyclopentadienides are also prepared by the similar methods. The main reagents have been the compounds of sodium, lithium or aluminium. C_5H_5Na is prepared by reacting freshly distilled cyclopentadiene with metallic sodium in tetrahydro furan.

Lithium alkyls are obtained by reacting pallets of lithium with appropriate organic halide in a hydrocarbon solvent such as pentane. The main reagent is now used to replace halogen atoms form a compounds of the desired metal.

$$2C_5H_5Na + FeCl_2 \rightarrow (C_5H_5)_2Fe + 2NaCl$$

Many compounds having general formula $M(C_5H_5)_2$ are being prepared by this method where M stands for one of the first row transition elements *i.e.,* from titanium to nickel.

Vinyl allyl, methallyl and isopropenyl lithium are also prepared by similar methods. Both phenyl and butyl lithium exchange rapidly with tetra vinyl lead and tetra vinyl tin in ether or in pentane at room temperature.

$$(CH_2 = CH)_2Sn + 4PhLi \rightarrow 4CH_2 = CHLi + Ph_4Sn$$
$$(CH_2 = CH)_4Pb + 4BuLi \rightarrow 4CH_2 = CHLi + Bu_4Pb$$

Vinyl lithium is also prepared from tetra vinyl-lead or tin and metallic lithium in diethyl ether using trace of benzophenone as catalyst.

$$(CH_2 = CH)_4Pb + 4Li \rightarrow 4CH_2 = CHLi + Pb$$

Similarly allyl and methallyl lithium are also prepared as follows :

$$(CH_2 = CHCH_2)_4 Sn + 4PhLi \xrightarrow[\text{ether}]{\text{Diethyl}} 4CH_2 = CH.CH_2Li + Ph_4Sn$$

(e) *Metallation reactions :* The reaction in which there occurs metal hydrogen exchange is called metallation reaction. The general reaction is

$$RH + MY \rightleftarrows RM + HY$$

It is exemplified by the following reactions :

$$C_5H_6 + CH_3Li \rightarrow CH_4 + C_5H_5Li$$
$$CH_3C \equiv CH + C_2H5Br \rightarrow C_3H_6 + CH_3C \equiv C.MgBr$$
$$n.C_4H^-_9\ Na^+ + C_6H_6 \rightarrow n.C_4H_{10} + 6H^-_5\ Na^+$$

The reaction of a metal with an organic halide is a convenient methods for the preparation of organometallics derived from reactive metals such as lithium, magnesium or zinc. Dialkyl ethers, particularly diethyl ether, provide an inert, slightly polar medium in which the organometallic compounds are usually soluble. Care is needed to exclude moisture and air, which would otherwise react with organometallic product. As we saw earlier, this is usually achieved by using an inert atmosphere of dry nitrogen, ensuring that all the apparatus that is used is dry

$$CH_3I + 2Li \xrightarrow{\text{ether}} \underset{\text{methyl - lithium}}{CH_3Li} + LiI$$

$$CH_3CH_2Br + Mg \xrightarrow{\text{ether}} \underset{\text{ethylmagnesium bromide}}{CH_3CH_2MgBr}$$

Chloro-bromo or iodo-alkanes can be used in these reactions, but fluoro alkanes do not react with either lithium or magnesium. After the organometallic reagent has been prepared, it is not isolated but rather used without further treatment perhaps, to decant the ether solution from any excess metal or sediment. Unlike Grignard reagents, alkyl-lithium compounds can be prepared in hydrocarbon solvent, for example benzene of hexane. This is an advantage because some oraganolithium compounds react with ethers and so cannot be prepared in ether solution.

Sodium represents a special problem because of the very high reactivity of organosodium compounds towards ether and organic halides. Again hydrocarbon solvents are used, but even so special preparative techniques are necessary in order to prevent the ionic organosodium compound reacting with the halide as fast as they are formed to produce alkanes (Wurtz reaction).

(f) *Metal halogen exchange reaction :* The method has been found to be useful where direct reaction of the metal with halogen does occur. The general reaction may be put as follow :

$$RH + MY \rightarrow RM + HY$$

n-Butyl- lithium is the most important compound for the preparation of organolithium compounds by involving metal halogen exchange. Lithium has been attached to more electronegative organic radical or the radical with would form the more stable carbanion. Hence p-dibromobenzene produces a good yield of p-bromophenyl lithium in ether.

$$p-C_6H_4Br_2 + Bu^nLi \xrightarrow{\text{Ether}} p-Br.C_6H_4Li + Bu^nBr$$

Generally in the reaction of a metal halide with an organometallic of another metal, the halogen ends up combined with the less electronegative metal and the organic group with the more electronegative metal.

The most versatile alkylating or arylating reagents in an exchange process with another metal halide are therefore the organometallic derivatives of the less electronegative metals. Thus we could predict that organolithium or Grignard reagents would be most useful. Because these reagents are readily available, they are in fact the most commonly used in the laboratory. Some sample reactions are :

$$CdCl_2 + 2C_2H_5MgBr \rightarrow (C_2H_5)_2Cd + MgCl_2 + MgBr_2$$

$$HgCl_2 + 2CH_3MgBr \rightarrow (CH_3)_2Hg + MgBr_2 + MgCl_2$$

$$ZnBr_2 + 2C_6H_5MgBr \rightarrow (C_6H_5)_2Zn + 2MgBr_2$$

$$SCl_4 + 4C_6H_5MgBr \rightarrow (C_6H_5)_4Si + 4MgBr_2$$

All these reactions involve Grignard reagents. Organolithium reagents are more reactive and so are used in cases where complete substitution of halogen by alkyl groups is required but cannot be achieved by using a Grignard reagent. For example :

$$TlCl_3 + 2RMgCl \rightarrow R_2TlCl + 2MgCl_2$$

$$TlCl_3 + 3RLi \rightarrow R_3Tl + 3LiCl$$

Whereas the Grignard reagent can replace only two of the three chlorine atoms of thallium (III) chloride, the lithium reagent will replace all three.

(g) *Simple ligand Exchange :* The general reaction may be put as follows :

$$A + MY \rightarrow AM + Y$$

where A and Y represent ligand, M for metal atom respectively.

(i) *Exchange of one neutral ligand for another :* Isonitrile replaces CO a follows :

$$Ni(CO)_4 + 4PhNC \rightarrow Ni(CNPh)_4 + 4CO \uparrow$$

$$Cr(CO)_6 + 3PhNC \rightarrow Cr(CO)_3(CPh)_3 + 3CO \uparrow$$

(ii) *Exchange of neutral ligands for anionic ligands :* Isocyanide, halocarbons and olefin complexes especially the platinum may be prepared by displacing anionic groups from a complex anion.

$$K_2[PtCl_4] + CO \rightarrow K[PtCOCl_3] + KCl$$

$$\downarrow CO$$

$$Pt(CO)_2Cl_2 + KCl$$

$$K_2[PtCl_4] + C_2H_4 \rightarrow K[Pt(C_2H_4)Cl_3] + KCl$$

METAL COMPLEXES OF ARENE TRANSITION

The majority of arene complexes are of the type $(C_6H_6)_2M$ where M = V, Cr, Mo, W, Mn, Te, Re, Fe, Rn, Os, Co, Rh and Ir. Arene complexes are those compounds in which the aromatic ring compounds such as benzene, act as ligand. The first such compound, di-arene chromiums, were obtained by Hein and coworkers, about 40 years ago, though their exact nature was established later by On sager and Zeiss, as well as Fischer and Hafner.

Preparation

Arene metal complexes are formed only with metals in vary low oxidation states. These can be prepared as :

(i) Fischer's Reducing Friedel Craft's Method

These are prepared the reduction of a suitable metal salt with aluminium powder followed by the addition of arene ligand to metal. The reaction has been catalysed by $AlCl_3$ and trace of mesitylene.

$$3CrCl_3 + 2Al + 6C_6H_6 + AlCl_3 \rightarrow 3[\pi - C_6H_6)_2Cr^+ AlCl_4]^-$$

The neutral complex is obtained by the disproportion of the chrominum cation by dilute alkali as

$$2\left[(\pi - C_6H_6)_2 Cr\right]^+ \xrightarrow{H_2O} (\pi - C_6H_6)_2 Cr + Cr^2 + 2C_6H_6$$

Using the above method, arene complexes of many of the transition metal may be obtained. Benzene can be replaced by toluene , mesitylene, hexamethyl benzene and xylene etc. Hence vanadium complex may be prepared as follows

$$VCl_4 + Al + Mesitylene \xrightarrow{AlCl_3} \left[(\pi - Mes.)_2 V\right]^+$$

$$3\left[(\pi - Mes.)_2 V\right]^+ \xrightarrow[\text{hydrolysis}]{\text{Strongly alkaline}} 2\left[(\pi - Mes.)_2 C\right] + V^{3+} + 2 Mes.$$

(ii) From Metal Carbonyls

Arene metal carbonyl complexes are obtained by the action of metal carbonyls with arenas in the presence of ultra violet light or high boiling diglyme.

$$C_6H_6 + Mo(CO)_6 \xrightarrow[\text{Ether}]{\text{U.V. or}} C_6H_6Mo(CO)_3 + 3CO\uparrow$$

$$C_6H_5.CH_3 + Mo(CO)_6 \rightarrow C_6H_6CH_2Mo(CO)_3 + 3CO\uparrow$$

The arene metal tricarbonyl complexes where the metal is chromium, molybdenum or tungsten are also prepared by this method. In some cases the reaction between metal carbonyls and the arene ligand occurs only in the presence of aluminium trialide which acts as a catalyst.

$$Co_2(CO)_8 + AlBr_3 + C_6H_6 \rightarrow [(C_6H_6)_3Co_3(CO)_2]^+$$

$$Hg\left[Co(CO)_4\right]_2 + AlCl_3 + C_6H_6 \xrightarrow{60-70^\circ C} \left[(C_6H_6)_3Co_3(CO)_2\right]^+$$

Properties

The neutral di π-arene complexes form well defined crystals which are insoluble in water but are soluble in organic solvents. They sublime in vacuum at ~ 100°C. The complexes are thermally stable upto about 300°C. dibenzene chromium is most stable of the derivatives. Most of the neutral di-arene metal complexes get oxidized readily by air or oxygen to give diarene cations which are very resistant towards further oxidation. The aromatic ligands, sometimes one and sometimes both, can be displaced rather from these arene complexes.

$$(C_6H_6)_2Cr + Cr(CO)_6 \rightarrow 2C_6H_6Cr(CO)_3$$

$$(C_6H_6)_2Cr + 6PF_3 \rightarrow 2C_6H_6 + Cr(PF_3)_6$$

Bis-arene chromium compounds are sufficiently good reducing agents to act as electron donors to strong-π acceptor acids such as tetracyano-ethylene.

$$(\pi - C_6H_6)_2Cr + (CN)_4C_2 \rightarrow [(\pi - C_6H_6)_2Cr]+[(CN)_4C_2]^-$$

STRUCTURE AND BONDING

There has been a lot of controversy regarding the structure of bis benzene chromium. There is a general agreement now that it is a sandwich type compound like ferrocene with the staggered arrangement of the two benzene rings with respect to each other.

X-ray and electron diffraction measurements on bisbenzene chromium and other arene complexes now show conclusively that all C – C bond lengths are equal.

Electron diffraction studies show that the di-benzene chromium has sandwich structure in which the C – C distances are all equal 1.423 ± 0.002Å, the C – H bond distance is 1.190 ± 0.005Å and the Cr – C, bond distance is 2.150 ± 0.002Å. The structure is shown as follows: The six P_{xz} electrons of benzene form six molecular orbitals which interacted with the metal orbitals by virtue of their same symmetry. In order to investigate more closely the nature of bonding, the relative ordering of energy levels is taken into consideration. The results as calculated by Schus Torovitch and Dyatkina are summarised in the following energy level diagram. The energy level diagram is different from that of the benzene in many respects. This is because the energies of the benzene molecular orbitals have been lower than those of the cyclo pentadienyl radical whereas the energies of the chromium atomic orbitals have been found to be higher than those of the iron.

Because of these differences in di π – benzene chromium the donor acceptor interactions make less contribution to the boning than in ferrocene and e_g bonds have much donor acceptor properties. Unlike the ferrocene e_{2g} electrons take part in the metal ring bonding. It favour a completely delocalized ring structure for the molecule. The above structure of the di π – benzene chromium has been found to be in close agreement with the known facts against the molecule.

LITHIUM ORGANOMETALLIC COMPOUNDS

The important organolithium compounds are alkyl or aryl lithium. These are prepared a follows :

By the Interaction of Alkyl Halides with Metals

Some organolithium compounds can be prepared by heating an alkyl or an aryl halide in ether or benzene with metallic lithium, *e.g.,*

$$RX + 2Li \rightarrow \underset{\text{Alkylithium}}{RLi} + LiX$$

$$CH_3I + 2Li \xrightarrow[\text{ether}]{-10^oC} \underset{\text{Mthyl lithium}}{CH_3Li} + LiI$$

$$\underset{\text{n - Butyl Chloride}}{CH_3CH_2CH_2CH_2Cl} + 2Li \rightarrow$$

$$\underset{\text{n - Butyl lithium}}{CH_3CH_2CH_2CH_2Li} + LiCL$$

Alkyl bromides and iodides usually undergo Wurtz reaction. Alkyl chloride, aryl bromides and aryl iodides give good yields of organolityium compounds. Low temperature (–10°C) is kept to minimise Wurtz reaction.

From Organomercury Compounds

It is possible to prepare organolithium compounds conveniently treating the lithium metal with organomercury compounds *e.g.,*

$$R_2Hg + 2Li \rightarrow 2RLi + HG$$

Properties

The alkyl lithiums are either colourless liquids or solids with low mewlting points. They are insoluble covalent compounds which may have partial ionic character according to the nature of the alkyl group. Thus, methyllithium is covalent while n-butyllithium has considerable degree of polarity (due to partial ionic character).

Generally, alkyl lithiums behave like Grignard reagents but are more reactive. Similar to Grignard's reagents, it is possible to use organolithium compounds for synthesising most of the types of organic compounds. Furthermore, organolithium compounds have been found to resemble very much with or ganomagnesium compounds in chemical properties except few differences. Thus, it becomes possible to discuss organolithium compounds under two headings:

(a) Those in which they are resembling Grignard reagents and

(b) Those in which they differ from Grignards regents.

(a) Resemblance with Grignard reagents.

PREPARATION OF HYDROCARBONS

Like Grignard reagents, organolithium compounds react with compounds containing active hydrogen to yield the corresponding hydrocarbons, viz.,

$$RLi + HOH \rightarrow R - H + LiOH$$

$$RLi + R'OH \rightarrow R - H + LiOR'$$

$$RLi + BrCH_3 \rightarrow R - CH_3 + LiBr$$

$$CH_3CH = CHLi + R'OH \rightarrow CH_3CH = CH_2 + LiOR;$$

$$CH \equiv CLi + R'OH \rightarrow CH \equiv CH + LiOR'$$

$$[ef.CH_3MgI + C_2H_5OH \rightarrow CH_4 + MgIOC_2H_5]$$

Preparation of Alcohols

It is possible to prepare primary, secondary and tertiary alcohols by the action of organolithium compounds on HCHO, CH_3CHO and ketone, respectively.

(a) Primary alcohols

$$H-\overset{\overset{O}{\|}}{C}-H + RLi \rightarrow H-\underset{R}{\overset{OLi}{C}}-H \xrightarrow{HOH} H-\underset{R}{\overset{OH}{C}}-H + LiOH$$

Primary alcohol

$$\left[\text{ef. } H-\overset{\overset{O}{\|}}{C}-H + CH_3MgI \rightarrow H-\underset{CH_3}{\overset{OMgI}{C}}-H \xrightarrow{HOH} H-\underset{CH_3}{\overset{OH}{C}}-H + MgIOH\right]$$

(b) Secondary alcohols

$$CH_3-\overset{\overset{O}{\|}}{C}-H + RLi \rightarrow CH_3-\underset{R}{\overset{OLi}{C}}-H \xrightarrow{HOH} CH_3-\underset{R}{\overset{OH}{C}}-H$$

Secondary alcohol

(c) Tertiary alcohols

$$CH_3-\overset{\overset{O}{\|}}{C}-R' + RLi \rightarrow CH_3-\underset{R}{\overset{OLi}{C}}-R' \xrightarrow{HOH} CH_3-\underset{R}{\overset{OH}{C}}-R'$$

Tertiary alcohol

3. Preparation of Ethers

It is possible to prepare higher ethers by the action of organolithium compounds ona monochloro ether.

$$ROCH_2Cl + R'Li \rightarrow ROCH_2R' + LiCl$$

Preparation of Amines

These may be prepared by the reaction between RLi and chloramine as follows.

$$RLi + ClNH_2 \rightarrow RNH_2 + LiCl$$

Preparation of Thioalcohols

These may be prepared by the action of organolithium compounds on sulphur and hydrolysing the product.

$$RLi + S \rightarrow RSLi \xrightarrow{HOH} RSH + LiOH$$

Preparation of Alkyl Iodides

These may be prepared by the action of iodine on RLi.

$$RLi + I_2 \rightarrow RI + LiI$$

Preparation of Aldehydes

It becomes possible to prepare aldehydes by the reaction between organolithium compounds with hydrogen cyanide.

$$HC \equiv N + RLi \rightarrow \underset{\substack{|\\ R}}{HC} = N.Li \xrightarrow[-LiOH]{HOH} RCH = NH \xrightarrow{HOH} RCHO + NH_3$$

Preparation of Ketones

Like aldehydes, ketones can be prepared by the reaction between organolithium compounds with alkyl cyanide.

$$CH_3C \equiv N + RLi \rightarrow \underset{\substack{|\\ R}}{CH_3.C} = NLi$$

$$\xrightarrow[-LiOH]{H_2O} CH_3.C = NH \xrightarrow{H_2O} \underset{\substack{|\\ R}}{CH_3.C} = O + NH_3$$

It is also possible to prepare ketones from carbon dioxide (see (b) type of reactions) or carboxylic acids.

Alkyl Cyanides

These may be prepared by the interaction of cyanogen chloride and RLi.

$$RLi + ClCN \rightarrow RCN + LiCl$$

Preparation of Other Organometallic Compounds

It is possible to prepare other organometallic compounds by the action of metal halide on organolithium compounds as follows.

$$3CH_3Li + AuBr_3 \rightarrow (CH_3)_3Au + 3LiBr$$

$$2CH_3Li + CdCl_2 \rightarrow (CH_3)_2Cd + 2LiBr$$

$$4CH_3Li + SiCl_4 \rightarrow (CH_3)_4Si + 4LiBr.$$

(b) Differences from Grignard Reagents

As organolithium compounds have been found to be more reactive than Grignard reagents, the former undergo some additional reactions as well as differ in some of the Grignard reactions, viz.

1. Addition on Carbon Dioxide (Preparation of Ketones)

Carbon dioxide forms carboxylic acids the Grignard reagents but ketones with the organolithium compounds. The difference is attributed to the fact that the organolithium compounds have been somewhat more strongly nucleophilic than Grignard reagents and hence react with the intermediate resonance stablilized carboxyalate anion. Because of this reason carboxylic acids also find use in synthesising ketone from the lithium compounds.

2. Addition on Olefinic Double Bonds

Unlike Grignard reagents, organolithium compounds add on ethylenic double bonds. The addition occurs at high pressure (100 – 500 atmospheres) on simple olefins, *e.g.*, ethylene;

$$C_4H_9Li + CH_2 = CH_2 \rightarrow C_4H_9CH_2 - CH_2Li$$

$$\xrightarrow{CH_2\,=\,CH_2} C_4H_9(CH_2)_4Li \xrightarrow{CH_2\,=\,CH_2} \text{etc.}$$

But at atmospheric pressure on conjugated olefins.

Organolithium compounds also add on > C = NR to yield amines, *e.g.*,

$$C_6H_5-\overset{H}{\overset{|}{C}}=N\;C_6H_5+RLi\rightarrow C_6H_5-\underset{R}{\underset{|}{\overset{H}{\overset{|}{C}}}}-\overset{Li}{\overset{|}{N}}-C_6H_5\xrightarrow{HOH}C_6H_5-\underset{R}{\underset{|}{\overset{H}{\overset{|}{C}}}}-NH.C_6H_5$$

3. Addition on Activated Carbon Atom of Aromatic System

Unlike Grignard's reagents, organolithium compounds add on activated carbon atom of the aromatic system which on hydrolysis yields substitution products, *e.g.*, pyridine.

Structure of Organolithium Compounds

In the crystals of methyl-(Fig. 5.36 and ethyl-lithium (EtLi, m.p. 90) the lithium atoms are at the corners of a tetrahedron with alkyl groups centered over the facial planes. Although, the CH_3 groups is symmetrically bound to three Li atoms, the α–carbon of the C_2H_5 group is closer to one Li atom than the other two.

The alkyl bridge bonding is of the electron deficient multicentre type found in Be and Al alkyls and in boranes. Aggregate formation is due to Li – C – Li rather than to Li – Li bonding interactions. The electronic structure can be accommodated by MO theory. The question whether any significant direct Li – i bonding occurs in these tetrahedral remains controversial, MO calculations have suggested that there is bonding, but Raman and nmr spectra suggest that there is none.

USES OF ORGANOMETALLIC COMPOUNDS

Organometallic compounds are of immense importance.

(i) Tetaethyl lead, Et_4Pb, is the most important organometallic compound which is added in small proportion to petrol an antiknocking agent; such petrol is termed as leaded petrol or ethyl petrol.

One part of TEl per thousand parts of petrol is used along with a small amount of ethylene bromide to prevent lead from being deposited in the cylinder head. The ethylene bromide converts lead in to lead dibromide which is sufficiently volatile and therefore can be removed with the exhaust gases.

(ii) Many organometallic compounds find use as catalysts in polymerisation reactions, *e.g.*,

(a) Butylsodium for converting butadiene to buna rubber.

(b) Alkyls of lithium for converting alkenes and dienes to elastomers.

(c) Trialkyl aluminium (Ziegler-Natta type catalysts) for the low temperature polymerisation of alkenes.

(iii) The π-complex of ethylene with $PdCl_2$ finds use to produce acetaldehyde which can then get converted into vinyl acetate.

The complex gets hydrolysed by water in presence of $CuCl_2$ to yield acetaldehyde, palladium chloride and precipitated CuCl. The latter then gets converted back to $CuCl_2$.

(iv) Alkyl cobalt carbonyls get formed in the conversion of alkenes into alcohols (which find use in detergent manufacturing) in the oxo- or hydroformylation reaction, by the action of hydrogen and carbon monoxide in the presence of $Co_2(CO)_8$.

$$Co_2(CO)_8 + H_2 \xrightarrow[\text{and pressure}]{\text{high temp.}} 2HCo(CO)_4$$

$$R.CH = CH_2 + H\ Co(CO)_4 \rightarrow R.CH_2CH_2.Co(CO)_4$$

$$\uparrow \qquad\qquad \downarrow CO + H_2$$

$$HCo(CO)_4 + R.CH_2.CHO$$

$$\downarrow H_2$$

$$\text{Detergents} \leftarrow R.CH_2CH_2CH_2OH$$

(v) There are many orgnometallic compounds which are medicinally important, *e.g.*, mercurochrome (antiseptic), salvarsan and neosalvarsan are used against syphilis and other disease.

TELLURIUM

In 1782 Muller Von Reichenstein discovered the element es. Klaproth named the element as tellurium (in Latin, tell us means earth). Berzelius showed its analogy with sulphur and selenium.

Occurrence

It occurs in native state in minute quantities in sulphur deposits. Generally it is found in combination with metals as tellurides. Some important telluride ores are :

Sylvanite or Graphite tellurium	$(Ag, Cu)\ Te_2$
Nagyagite or Black tellurium	$(Au, Pb)_2\ (Te, S, Sb)_2$
Tetradymite	Bi_2Te_3
Petzite	$(Au, Ag)_2Te$
Calaverite	$(Au, Ag)Te_2Au,Te$
Coloradoite	HgTe
Tellurite	TeO_2

The chief sources of tellurium at present are the slime from electrolytic tanks of copper and lead refineries and the flue dust from smelters using telluride of gold ores.

Extraction

Tellurium is also obtained from copper refineries in another way. The anodic mud obtained during the electrolytic refining of copper or the residue which escapes roasting is fused with sodium nitrate (an oxidising agent) and sodium hydroxide when selenium and tellurium form selenite and tellurite respectively.

$$Se + 2NaOH + 2NaNO_3 \rightarrow Na_2SeO_3 + 2NaNO_2 + H_2O$$

$$Te + 2NaOH + 2NaNO_3 \rightarrow Na_2TeO_3 + 2NaNO_2 + H_2O$$

The fused mass is extracted with water and then acidified with HCl when tellurium is precipitated out as oxide while selenium remains in solutions as selenious acid.

$$Na_2SeO_3 + 2HCl \rightarrow H_2SeO_3 + 2NaCl$$

$$Na_2TeO_3 + 2HCl \rightarrow H_2TeO_3 + 2NaCl$$

Sulphur dioxide is passed through the filtrate to get selenium. Tellurium dioxide is reduced by heating it with charcoal.

From Tetradymite

The finely powdered ore is made into a stiff paste with sodium or potassium carbonate in olive oil. The paste is heated in a closed crucible to white heat when arsenic and antimony, present as impurities, volatilise off leaving behind, S, Se and Te in the residual mass. The latter is pulverised and dissolved in water. Now a current of air is passed through the filtrate to precipitate out tellurium while sulphur, selenium and a part of tellurium remain in solution. After filtering off, tellurium is purified by distillation in a current of hydrogen when gold, iron, manganese ad copper and separated. The filtrate after removing tellurium is acidified with HCl to precipitate out the residual tellurium. The complete extraction from tetradymite in the form of flow of flow sheet is given below :

The reactions involved in the extraction are noted here.

$$Bi_2Te_3 + 3Na_2CO_3 \rightarrow 3Na_2Te + Bi_2O_3 + 3CO_3$$

$$Na_2Te + 2H_2 \rightarrow 2NaOH + H_2Te$$

$$2H_2Te + O_2 \rightarrow 2H_2O + 2Te$$

From Sylvanite : (ag, Au) Te_2

This mineral contains about 35% of tellurium, 46-61% silver and 1-18% gold. The powdered ore is heated with HCl in which arsenic and antimony together with other metals, dissolve. The residue is dissolved in aquaregia and the excess of acid is evaporated. From the solution gold is precipitated by $FeSO_4$ and from the filtrate tellurium is precipitated by passing SO_2.

Tellurium can be extracted from sylvanite by another method. In this method, the powdered ore is heated in a current of chlorine when silver forms the non-volatile silver chloride while tellurium forms the volatile tetrachloride. The volatile tellurium tetrachloride accompanied by sulphur, selenium and antimony chloride, i dissolved in dilute HCl and tellurium is precipitated by sodium bisulphite. Tellurium so obtained may be further purified by distillation in hydrogen.

Preparation of Tellurium

Free tellurium is prepared from the precipitated TeO_2 by either of the following processes :

(i) Dry Process

TeO_2 is dried and mixed with powdered coke or charcoal. The mixture is heated to a temperature of 500. The yield is upto 95%.

$$TeO_2 + 2C \xrightarrow{450-500^o} Te + 2CO$$

(ii) Wet Process

In wet process, TeO_2 is dissolved in hot conc. HCl. The acidic solution is diluted with water taking care not to precipitate TeO_2.SO_2 is passed through the solution when tellurium precipitates as a dark grey powder. This is filtered out, washed and dried.

$$TeO_2 + 2SO_2 + 2H_2O \rightarrow Te + 2H_2SO_4$$

Purification

The element as obtained by above methods is not pure. It is purified by distillation in a current of O_2 or by treating with potassium cyanide. KCN gives K_2Te.

$$2K_2Te + 2H_2O + O_2 \rightarrow 4KOH + 2Te$$

Properties of Tellurium

Physical

Te is a brittle, silvery-grey metallic solid existing in several allotropic forms. The most familiar allotropes are the stable grey crystalline form and the amorphous form.

1. Crystalline Form

This form is obtained by heating amorphous tellurium.

(i) Crystaline tellurium is silvery-white in colour with hexagonal crystals and has a metallic lustre. (ii) It is brittle and can be powdered easily. (iii) It is a poor conductor of electricity and heat. The electrical conductivity of Te varies very slightly with change in illumination, but there is a marked increase in conductivity by adding small quantities of foreign metals. (iv) Crystalline Te has specific gravity 6.25, 6.31 depending upon the method of preparation. (v) It metals at 453°C and boils at 1400°C and gives out golden yellow coloured vapours. (vi) The crystalline form is insoluble in water and CS_2 but soluble in HNO_3. (vii) The vapour density at 1400°C corresponds to Te_2 *i.e.,* the molecules are diatomic.

2. Amorphous Form

The amorphous form is obtained by the reduction of tellurous or telluric acid by SO_2, hydrazine, or hydroxylamine.

(i) It is grey powder of specific gravity 0.015. (ii) It melts at 444.8 and boils at 1390°C giving a golden yellow vapour. (iii) The vapour density corresponds to formula Te_2 showing the molecules to be diatomic.

3. Colloidal Tellurium

This is obtained by the reduction of telluric acid or by electrolysis of a solution of tellurium in nitric acid. The colour of colloidal solution is dark brown.

Chemical Properties

1. Reaction with Oxygen

Te is stable in air at ordinary temperature by forms the dioxide when hated in air or burnt in oxygen with a bright blue flame, green at edge.

$$Te + O_2 \rightarrow TeO_2$$

2. Reaction with Ozone

Ozone with water oxidases tellurium to telluric acid.

$$Te + H_2O + O_3 \rightarrow H_2TeO_4$$

3. Reaction with Hydrogen

Hydrogen combines with Te directly to form hydrogen telluride (H_2Te). The reaction with nascent hydrogen occurs at a very low temperature.

4. Reaction with Water

Freshly precipitated Te begins to react with water at 50°C to form TeO_2 with evolving hydrogen.

$$Te + 2H_2O \xrightarrow{50^o} TeO_2 + 2H_2$$

With crystalline form of tellurium, the reaction occurs at 100°C.

5. Reaction with Halogens

Fluorine reacts with Te forming tetrafluoride while at low temperature –78° a hexafluoride is formed.

$$Te + 2F_2 \rightarrow TeF_4$$

$$Te + 2F_2 \xrightarrow{-78^o} TeF_6$$

Tellurium forms di and tetra chlorides with chlorine in cold.

$$Te + Cl_2 \rightarrow TeCl_2$$

$$Te + 2Cl_2 \rightarrow TeCl_4$$

Bromine combines with Te at ordinary temperature giving tellurium dibromide and tellurium tetrabromide.

$$Te + Br_2 \rightarrow TeBr_2$$

$$Te + 2Br_2 \rightarrow TeBr_4$$

I_2 and Te do not react directly at ordinary temperature, but when melted together TeI_2 is obtained.

6. Reaction with Non-Metals

Non-metal like sulphur react with Te forming sulphides. Phosphorus forms tellurium phosphide. Arsenic forms arsenic telluride and Sb form antimony telluride when hydrides of these react with Te.

7. Reaction with Metals

Te combines with sodium to form dark red alkali poly tellurides.

$$Te + 2Na \rightarrow Na_2Te$$

Molten Mg reacts vigorously with Te.

$$Mg + Te \rightarrow MgTe$$

Reaction of aluminium with Te yields aluminium telluride. Zn and lead form alloy of Te which is widely used.

8. Reaction with Acids

Te reacts with con. H_2SO_4 with the liberation of SO_2 while HCl does not react.

$$Te + 2H_2SO_4 \rightarrow TeSO_4 + SO_2 + 2H_2O$$

Tellurium is oxidised by conc. HNO_3 to tellurous acid (H_2TeO_3).

$$Te + 2HNO_3 \rightarrow TeO_2 + NO + NO_2 + H_2O$$

$$TeO_2 + H_2O \rightarrow H_2TeO_3$$

9. Reaction with Alkalies

Te dissolves in hot alkalies forming tellurides and tellurites.

$$2NaOH + Te + H_2O \rightarrow Na_2TeO_3 + 2H_2$$

$$4NaOH + 2Te \rightarrow 4Na_2TeO + 2H_2O + O_2$$

10. Reaction with $AuCl_3$

Tellurium tetra-chloride is obtained by reacting $AuCl_3$ with tellurium.

$$4AuCl_3 + 3Te \rightarrow 3TeCl_4 + 4Au$$

Uses of Tellurium

1. Te is used in making alloys as lead tellurium and aluminium alloys. Te in parts hardness, toughness and increased corrosion resistance to lead even in an amount of 0.02 to 0.25%.
2. It is used as a delicate test for sterilization in bacteriology.
3. Tellurium has been employed for toning prints in photographic works especially in silver prints.
4. Te compounds have been used as dye stuffs.

5. Colloidal tellurium is used as a fungicide.
6. Like selenium compounds, a few compound of tellurium have antiknock properties. Diethyl telluride, $Te(C_2H_5)_2$ is one of such compounds used as an antiknock.
7. Tellurium has been suggested in the treatment of cancer and syphilis.
8. Tellurium can be used as an indicator electrode for potentiometric titration even in non-aqueous solvents.

COMPOUNDS OF TELLURIUM

The compounds of tellurium resemble to those of sulphur and selenium closely, but it possesses more metalic character than these elements.

1. Tellurium Hydride

It is prepared by the combination of hydrogen and tellurium.

$$H_2 + Te \rightarrow H_2Te \text{ (By distillation of Te in } H_2\text{)}$$

It can also be obtained by treating a telluride with dil HCl.

$$K_2Te + 2HCl \rightarrow H_2Te + 2KCl$$

$$ZnTe + 2HCl \rightarrow H_2Te + ZnCl_2$$

Properties

(i) It is a colourless gas with an unpleasant colour like H_2S. It is less stable then H_2S but pure H_2Te is fairly stable in dark. It decomposes when exposed to light especially in presence of moisture.

$$H_2Te \rightarrow H_2 + Te$$

(ii) The gas H_2Te burns in air with a pale-blue flame.

$$2H_2Te + 3O_2 \rightarrow 2H_2O + 2TeO_2$$

(iii) The solution of H_2Te in water oxidises in air and becomes red due to separation of tellurium.

(iv) It serves as reducing agent and reduces the $FeCl_3$, $HgCl_2$ solution.

$$2FeCl_3 + H_2Te \rightarrow 2FeCl_2 + 2HCl + Te$$

$$2HgCl_2 + H_2Te \rightarrow 2Hg_2Cl_2 + 2HCl + Te$$

(v) Hydrogen telluride reacts with chlorine in the following way.

$$H_2Te + Cl_2 \rightarrow Te + 2HCl$$

$$Te + Cl_2 \rightarrow TeCl_2$$

2. Oxides of Tellurium

It forms the following oxides :

(a) Tellurium Monoxide

It is obtained as black solid on heating sulphite.

$$TeCO_3 \rightarrow TeO + SO_2$$

(b) Tellurium Trioxide

It is obtained as an orange-yellow powder by heating telluric acid to redness. It is insoluble in water.

On heating it dissociates into TeO_2 and O_2.

(c) Tellurium Dioxide

It is formed by burning tellurium in oxygen.

$$Te + O_2 \rightarrow TeO_2$$

Properties

It is a white solid in soluble in water. It is weak amphoteric substance. It dissolves in alkalies forming tellurites.

$$2KOH + TeO_2 \rightarrow K_2TeO_3 + H_2O$$

From this a hydrated form is precipitated by adding acid.

$$K_2TeO_3 + 2HCl \rightarrow H_2TeO_3 + 3KCl]$$

(i) With acids like HNO_3, TeO_2 forms a basic nitrate, $2TeO_2.HNO_3$ or $Te_2O_3(OH)NO_3$.

With H_2SO_4 it gives tellurium sulphate.

Tellurous acid, H_2TeO_3 (1) This is obtained as a precipitate when a solution of TeO_2 in nitric acid is poured in water.

$$TeO_2 + H_2O \xrightarrow{\text{Hydration}} H_2TeO_3$$

(2) By treating a tellurite with acid.

$$K_2TeO_3 + 2HCl \rightarrow 2KCl + 2TeO_3$$

Properties

(i) It is a white amorphous substance. (ii) It is sparingly soluble in water. (iii) It is soluble in alkalies and gives tellurites which are usually more complex.

$$H_2TeO_3 + 2KOH \rightarrow K_2TeO_3 + 2H_2O]$$

Telluric acid- $H_2TeO_4.2H_2O$

Preparation

1. It is prepared by the oxidation of tellurons acid by CrO_3 or $KClO_3$.

$$H_2TeO_3 + O + 2H_2O \rightarrow H_2TeO_4.2H_2O$$

2. By the action of HCl on tellurate.

$$K_2TeO_4 + 2HCl \rightarrow H_2TeO_4 + 2KCl.$$

Properties

(i) It is a white solid.

(ii) It is sparingly soluble in cold, readily in hot water and much less ionised.

(iii) It loses water when heated beyond 140°C and gives trioxide.

$$H_2TeO_4 \xrightarrow{150^\circ C} H_2O + TeO_3$$

(iv) It is reduced to the element by SO_2 and other reducing agents.

(v) It forms various hydrates and undergoes polymerisation readily.

(vi) It forms colloidal or semi-colloidal substance and crystalliizes with salts such as phosphates, arsenites, oxalates, etc.

HALOGEN COMPOUNDS

Tellurium Hexafluoride-TeF_6

Preparation

When fluorine acts at – 78°C, it forms tellurium hexafluoride TeF_6.

$$Te + 3F_2 \xrightarrow{-78^\circ C} TeF_6$$

Properties

(i) It is a colourless gas. (ii) Its m.p. is –36°C and b.p. – 25.5°C. (iii) It is white crystalline at – 36°C. (iv) Unlike SF_6 and SeF_6 it is completely hydrolysed by water.

$$TeF_6 + 6H_2O \rightarrow H_6TeO_6 + 6HF$$

This is because of the large size of tellurium by which it can higher coordination number essential in the first stage of hydrolysis. TeF_6 can also add F^- ions to forms $[TeF_7]^-$ and $[TeF_8]^{2-}$ species.

Tellurium tetrafluoride, TeF_4 is obtained by reaction of TeO_2 with SeF4 at 0°C followed by the evaporation of the excess reagent and the $SeOF_2$.

$$TeO_2 + 2SeF_4 \rightarrow TeF_4 + 2SeOF_2$$

It is a colourless solid.

Tellurium Tetrachloride, TeCl4

Preparation

(i) It is obtained when excess of chlorine reacts with heated Te.

$$Te + 2Cl_2 \rightarrow TeCL_4$$

(ii) It is best formed by the action of S_2Cl_2 sulphur monochloride on tellurium.

$$Te + 2S_2Cl_2 \rightarrow TeCl_4 + 4S$$

Properties

(i) It is a white solid, m.p. 224°C and b.p. 414°C.

(ii) It is extremely hydroscopic and hydrolysed by water.

$$TeCl_4 + 2H_2O \rightarrow TeO_2 + 2HCl$$

(iii) $TeCl_4$ decomposes on heating beyond 530°C.

$$TeCl_4 \xrightarrow{530^\circ C} TeCl_2 + Cl_2$$

(iv) On heating $TeCl_4$ with Te, dichloride is formed.

$$TeCl_4 + Te \xrightarrow{Heat} 2TeCl_2$$

(v) $TeCl_4$ dissolves in HCl forming H_2TeCl_6

$$2HCl + TeCl_4 \rightarrow H_2TeCl_6$$

Dihalides

$TeCl_2$ (a black solid) and $TeBr_2$ (a green solid) are obtained by the direct union of elements in suitable proportions. $TeCl_2$ is also formed when $TeCl_4$ and Te are heated together while $TeBr_2$ is formed by heating the tetrabromide.

$$TeCl_4 + Te \rightarrow 2TeCl_2$$

$$TeBr_4 \rightarrow TeBr_2 + Br_2$$

They are readily hydrolysed according to the equation

$$2TeX_2 + 3H_2O \rightarrow Te + H_2TeO_3 + 4HX$$

Both halides form bent molecules (Tetrahedral structure with two positions occupied by two lone pairs). The structural parameters for the molecules are:

d(Te – Cl) = 2.36 Å $\angle$ClTeCl = 150 – 180°

d(Te – Br) = 2.51 Å, $\angle$BrTeBr = 98°

SOME REDOX TITRANTS

1. Potassium Bromate

Bromate ion is nearly as powerful oxidant as permanganate in acidic medium. Standard solution of the reagent can be prepared directly from primary standard $KBrO_3$. This is stable for indefinite period.

Reaction of Potassium Bromate

Reducing agent converts first bromate to bromide and then with excess of bromate, free bromine is liberated.

$$BrO^-_3 + 6H^+ + 6e^- \rightarrow Br^- + 3H_2O$$

$$BrO^-_3 + 5Br^- + 6H^+ \rightarrow 3Br_2 + 3H_2O$$

Usually bromide is added to the test solution before the titration or it is included in the standard bromate solution so that only second reaction is involved. Oxidising action of bromate was first noticed by Balard but the first application of bromate as a titrimetric reagent was made by Koppeschaar.

Detection of End Point

End point in the titrations with $KBrO_3$ is determined by yellow colour of free bromate obtained after the completion of reaction. But Gyory employed methyl orange as indicator in direct titrations with potassium bromate. Various other dyestuffs like indigo, carmine, bordeaux, rhodamine B, fluorescein, etc. have been recommended for the location of end point. Titrations with bromate solution should be carried out slowly so that indicator change which is a time reaction may be readily detected.

PREPARATION OF STANDARD SOLUTION

Potassium bromate is a standard substance and can be easily obtained in a pure state by recrystallisation from water. To prepare a standard solution dry some finely powdered A.R. potassium bromate 1 – 2 hours at 120°C and allow to cool in a desiccator. Now weigh and dissolve in distilled water.

Applications of Bromate in Titrimetric Analysis

1. Determinations of Metals

Gyory determined arsenic (III) and antimony (III) with potassium bromate in a medium upto 1N hydrochloric acid using methyl orange as indicator.

$$2KBrO_3 + 3M_2O_3 + 2HCl \rightarrow 2KCl + 3M_2O_5 + 2HBr$$

(M = As or Sb)

In a similar way potassium bromate has also been applied in the determination of copper (I), thallium (I). selenium (IV), iron (II), vanadium (IV), mercury (II) and uranium (IV).

2. Determination of Metals of Oxinate

Various metals like aluminium, magnesium, calcium, thorium, titanium, iron (III), nickel, cobalt, gallium, zirconium, tungstate, molybdate, etc. are determined with bromate after precipitating them with oxine under specific conditions.

3. Determination of Hydrogen Peroxide

Rupp and Siebler have reported the indirect determination of hydrogen peroxide with potassium bromate.

4. Determination of Caro's Acid, Perdisulphuric Acid, Hydrogen Peroxide in Presence of Each Other

Glue found that Caro's acid is quantitatively reduced by bromate in slightly acidic medium.

Hydrogen peroxide and perdisulphuric acid are not attacked under the conditions given.

After the back titration te peroxide can be titrated with permanganate in the presence of manganous sulphate. Finally persulphate is determined by reducing with bromate.

5. Carbon Monoxide

Winkler employed the bromate titration of arsenious oxide for the semimicro determination of carbon monoxide in air.

6. Halides

Berg gave a method by which iodide, bromide and chloride can be titrated with potassium bromate in a medium of hydrocyanic acid.

$$BrO^-_3 + 3I^- + 3CN^- + 6H^+ \rightarrow 3ICN + Br^- + 3H_2O$$

$$BrO^-_3 + 3Br^- + 3CN^- + 6H^+ \rightarrow 3BrCN + 3H_2O$$

$$BrO^-_3 + 3Cl^- + 5CN^- + 6H^+ \rightarrow 2ClCN + 3BrCN + 3H_2O$$

7. Sulphur Compounds

The determination of thiocyanate, suphide, sulphite and thiosulphate has been studied with bromate by several workers.

8. Nitrogen Compounds

Hydrazine may be directly determined with bromate in sufficient concentration of hydrochloric acid using indigo as indicator.

$$3N_2H_4 + 2HBrO_3 \rightarrow 2HBr + 3N_2 + 6H_2O$$

Bromate oxidises hydroxylamine quantitatively to nitric acid.

$$BrO^-_3 + NH_2OH_3 \rightarrow Br^- + H_2O + NO^-_3 + H^+$$

An excess of bromate is added and the back titration is done iodometrically.

9. Organic Compounds

Determination of several organic compounds like oxalic acid, formic acid, salicylic acid, phenols, vitamin C, aniline, esters, glycerine, etc. have been described with bromate.

10. Determination of Double Bonds

Solutions of bromine and of bromide-bromate mixture have been widely used for the determination of unsaturation.

POTASSIUM IODATE

Potassium iodate is a strong oxidising agent. The reaction between iodate and powerful reducing agent such as potassium iodide in fairly acidic medium stops when the iodate is reduced to iodine.

$$IO^-_3 + 5I^- + 6H^+ \rightarrow 3I_2 + 3H_2O$$

In a slightly acidic medium of hydrochloric acid, the mode of reaction is entirely different, *i.e.,* iodate is reduced ultimately to ICl.

$$IO^-_3 + 6H^+ + Cl^- + 4e \rightleftarrows ICl + 3H_2O$$

In the presence of an excess of hydrochloric acid solution, ICl forms complex ion.

$$ICl + Cl^- \rightleftarrows ICl^-_2$$

This is most important of the reactions of potassium iodate, as under such conditions of acidity (3M – 9M) that the reagent is very powerful oxidising agent with a redox potential of +1.23 volts. Oxidation of iodate in a large excess of hydrochloric acid proceeds through various stages:

$$IO^-_3 + 6H^+ + 6e^- \rightarrow I^- + 3H_2O$$
$$IO^-_3 + 5I^- + 6H^+ \rightleftarrows 3I_2 + 3H_2O$$
$$IO^-_3 + 2I_2 + 6H^+ \rightleftarrows 5I^+ + 3H_2O$$

The above reactions may be summarised as follows :

$$IO^-_3 + 6H^+ + 4e^- \rightleftarrows I^+ + 3H_2O$$

Hence under these particular conditions, the equivalent weight is one fourth of the molecular weight 214.4/4 = 53.5.

Location of the End Point

Involving Extraction Solvent : The end point in iodate titrations can be located by adding an immiscible solvent such as chloroform and carbon tetrachloride to the aqueous test sample. The point is noted by the disappearance of the last traces of violet colour, due to iodine, from the solvent. The end point is extremely sharp. Iodine mono-chloride (yellow) is not extracted from the aqueous phase.

DYE STUFFS AND INDICATORS

Smith and Wilcox replaced the immiscible solvents by certain dye stuff as irreversible indicators, *e.g.*, Amarnath (C.I. 184), Briliant ponceaux 5R etc. Belcher and Clark reported the use of p-ethoxychrysoidine as a satisfactory reversible indicator.

Preparation of 0.1N Potassium Iodate

Dry A. R. sample of potassium iodate for one hour and then cool it in a desiccator. Weight out 5.350 g of finely powdered potassium iodate in water and make up the volume to one litre.

Applications of iodate in Titrimetric Analysis

1. Determination of Iodides

The determination of iodides with iodate involving an extracting solvent is based upon the reaction.

$$IO^-_3 + 6H^+ + 2I^- + 3Cl^- \rightarrow 3ICl^- + 3H_2O$$

$$1 \text{ ml. of } 0.05M - KIO3 \equiv 0.006347 \text{ g of } I^-$$

2. Determination of Metals

Direct Methods : A number of metals like arsenic (III), antimony (III), tin (II), thallium (I), etc. may be directly determined with potassium iodate.

$$As_2O_3 + KIO_3 + 2HCl \rightarrow As_2O_5 + ICl + KCl + H_2O$$

$$Sb_2O_3 + KIO_3 + 2HCl \rightarrow Sb_2O_5 + ICl + KCl + H_2O$$

$$2SnCl_2 + KIO_3 + HCl \rightarrow 2SnCl_4 + KCl + ICl + 3H_2O$$

$$2TlCl + KIO_3 + 6KCl \rightarrow 2TlCl_3 + KCl + ICl + 3H_2O$$

3. Indirect Methods

Peroxides of lead, manganese and barium can be indirectly determined with potassium iodate.

Vanadates after reducing tote quadrivalent state by potassium iodide in presence of hydrochloric acid may be determined by back titrating with iodate.

$$H_3VO_4 + HI + 2HCl \rightarrow VOCl_2 + 1/2I_2 + 3H_2O$$

$$2I_2 + HIO_3 + 5HCl \rightarrow 5ICl + 3H_2O$$

$$2HI + HIO_3 + 3HCl \rightarrow 3ICl + 3H_2O$$

4. Precipitation Methods

Copper (II) may be determined with iodate after reducing to copper (I) and precipitation as thiocyanate. Likewise Hg(I) may be determined with iodate after precipitation as chloride.

5. Determination of Carbon

Kihu and co-workers estimated carbon in graphite, acitvated carbon or charcoal be decomposing with potassium iodate and phosphoric acid at a temperature above 250°C.

6. Other Determinations

Potassium iodate has also has also been applied in the determination of thiosulphate, thiocyanate, hydrazine, sulphites, sodium hypochlorite, etc. Deshmukh *et al.*, have studied the oxidation of ascorbic acid with iodate.

3. VANADATES

Compounds of pentavalent vanadium are distinguished by their great oxidising power. This property of vanadate (V) has been utilized in titrimetric analysis. Many inorganic substances can be oxidized with this standard solution.

The normal redox potential of the system

$$VO^-_3 + 4H^+ + e^- \rightleftharpoons VO^{2+} + 2H_2O$$

is + 1.00V. The redox potential increases with increasing acid concentration.

Most of the determination are carried out in H_2SO_4, HCl or CH_3COOH solution at room or elevated temperature. Some reactions have to be catalytically accelerated because they proceed very slowly even at high temperature.

Syrokomskii and stiepin

(1935) first introduced vanadium (V) as a titrant for ferrous iron. Lange *et al.* gave a method for the determination of molybdenum with ammonium vanadate.

PREPARATION OF SOLUTION

Ammonium vanadate and sodium metavanadate can both be employed as titrants. Recrystallized and dried ammonium vanadate is said to be a primaiy standard, but the solution is generally standardised before use. A 0.1n solution is prepared by weighing out about 1.2 g of the salt and dissolving it in 80 ml of 1 : 1 sulphuric acid before dilution to one litre. The solution is stable for several months.

Standardisation of Solution

Vanadate solutions are commonly standardised against a standard solution of Mohr's salt and indirectly against dichromate employing N-phenylanthranilic acid solution as indictor.

Indicator

N-phenylanthranilic acid is commonly used a indicator in visual titrations with vanadium (V), diphenyl amine, diphenylbenzidine, diphenylamine p-sulphonic acid, etc. are often used as indicators.

Most of the compounds can also be titrated potentiometrically.

Analysis Involving Vanadate Solution

The titration of ferrous iron with vanadate is suitable in 13.5 N sulphuric acid medium when N-phenyl anthranilic acid is used as reversible indicator. Platinum and palladium may be determined by reduction with excess of standard ferrous and back titration with a vanadate solution.

Molybdenum

Pentavalent molybdenum can be easily titrated with vanadate solution, using N phenyl anthranilic acid as indicator.

Uranium

Several methods have been described for the determination of uranium after reduction to the quadrivalent state by titration with vanadate. Rao *et al*. used diphenyl benzidine a indicator but found it necessary to catalyse the reaction with oxalic acid.

2. Sulphur Compounds

Rao and Rao have given a method for the determination of sulphite in which oxidation to sulphate is done with vanadate solution in the presence of iodine monochloride as catalyst.

The determination of thiosulphate has been described with vanadate solution if the acidity is above 2.7 N.

3. Hypophosphite and Phosphite

Rao et al. reported that both phosphite and hypophosphite may be determined by oxidising with vanadate at higher temperature in the presence of catalyst.

4. Hydroquinone

Vanadate oxidises hydroquinone quantitatively in the presence of oxalic acid as catalyst. The concentration of hydroquinone should be less them 0.01M and acid concentration above 1M.

5. Organic Acids

A number of organic acids like oxalic, tartaric and formic acids may be determined with vanadate solution.

6. Indigo

Vanadate is said to be a better titrant than permanganate for the determination of indigo using oxalic acid as a catalyst.

7. Other Determinations

Singh and Singh (1954) employed vanadate as titrant in strong hydrochloric acid medium (7 – 7.5N) in the presence of iodine monochloride. A number of substance like hydrogen peroxide, hydrazine, arsenite, copper, lead dioxide, etc. may be determined.

CERIC SALTS

The so called ceric sulphate or sulphato cerate was first used as oxidising agent by Lange. It was due to Willard and Young which has placed the salt among the important titrimetric analytical reagent. In acid solutions, Ce(IV) is a strong oxidant. The acidity and the nature of the acid present affects the redox potential of the system (Ceric-cerous) and makes it certain that infact the cerium in present in these solutions in the form of various complex anions. The redox potential is 1.44 – 1.42 volts in H_2SO_4 (IN – 8N).

$$Ce^{4+} + e \rightleftarrows Ce^{3+}$$

The advantages of ceric sulphate as a standard oxidising agent are:

(1) The solution of ceric sulphate is remarkably stable for years.

(2) In contrast of permanganate it may be employed in the determination of reducing agents in the presence of high concentration of hydrochloric acid.

(3) The equivalent weight of a ceric salt i same as molecular weight. As the reaction in this case with reducing agents is as follows:

$$Ce^{4+} + e \rightleftarrows Ce^{3+}$$

While in this case of permanganate and dichromate, intermediate unstable valence states are formed according to experimental conditions.

(4) Ceric sulphate is a very versatile and elegant oxidising agent. It may be employed in most titrations in which permanganate has been used, and also for other determinations.

(5) The cerous ion is colourless.

Detection of the End Point

Certain titrations can be done without an indicator. It is better, however, to use an indicator when possible. Of the many indicators that have been used for ceric sulphate titrations some of them are : N-phenyl anthranilic acid, ferroin, 5–6 dimethyl ferroin and α α - dipyrudyl ferrous sulphate.

Preparation of Ceric Solution

Solution of ceric sulphate may be readily prepared by dissolving pure ceric sulphate or the more soluble ceric ammonium sulphate in dilute (1 – 2N) sulphuric acid. The solid reagents are relatively expensive. Ceric ammonium nitrate of technical quality, is inexpensive and is prepared after recrystallisation from hot dilute nitric acid (1.3) in the presence of a little ammonium nitrate followed by drying at 80 – 85°C.

Standardisation of Ceric Solution

(a) With Arsenic Trioxide

The simplest standardisation of ceric sulphate is against arsenic trioxide. The reaction is very slow, it is necessary to add a trace of osmium tetroxide as catalyst. The arsenious oxide is dissolved in sodium hydroxide solution. The solution is acidified with dilute sulphuric acid, 2 drops of osmic acid solution added, followed by indicator (N-phenyl anthranilic acid). The colour change is form orange red to pale blue.

$$As_2O_3 + 4OH^- \rightarrow 2HAsO^{2-}{}_3 + H_2O$$

$$2Ce^{+4} + H_3AsO_3 + H_2O \rightarrow 2Ce^{+3} + H_3AsO_4 + 2H^+$$

(b) Standardisation Against Potassium Ferrocyanide

Classen used potassium ferrocyanide in 2N sulphate acid solution with ferrion as indicator for standardising ceric sulphate. Various other substances like sodium oxalate, ferrous iron, ad potassium iodide have been employed for standardising ceric solution.

Applications

1. Determination of Metals

Willard et al. showed that trivalent antimony in hydrochloric acid solution can be titrated with ceric sulphate involving ferroin indicator.

Likewise a number of metals like arsenic, mercury, thallium, tellurium tin, rhenium, chromium, vanadium, ferrous ion, etc. have also been estimated.

2. Nitrogen compounds–Azid

Hydrazoic acid is oxidised quantitatively to nitrogen by an excess of ceric ammonium nitrate.

$$2N^-_3 + 2Ce^{4+} \rightarrow 3N_2 + 2Ce^{+3}$$

Similarly hydroxylamine may be determined after oxidation with ceric sulphate.

$$2NH_2OH + 4Ce\,(SO_4)_2 \rightarrow N_2O + 2Ce_2\,(SO_4)_3 + H_2O + 2H_2SO_4$$

3. S. Phosphorus Compounds

Bernnart has described methods based for the determination of hypophosphite.

$$4Ce^4 + H_3PO_2 + 2H_2O \rightarrow H_3PO_4 + 4Ce^{3+} + 4H^+$$

$$2Ce^{4+} + H_3PO_3 + H_2O \rightarrow H_3PO_4 + 2Ce^{3+} + 2H^+$$

4. Hydrogen Peroxide

The reaction of hydrogen peroxide with ceric sulphate.

$$2Ce^4 + H_2O_2 \rightarrow 2Ce^{3+} + 2H^+ + O$$

was shown to be quantitative in potentiometric titrations by Furman *et al.*

5. Organic Compounds

A number of organic compounds like fatty acids, hydroxy compounds, carbohydrates, unsaturated, compounds and oxine have been estimated against ceric sulphate.

PREPARATION OF TITANOUS SALTS

Titanous, *i.e.,* titanium (III) salts are energetic reductants but have the disadvantage that they are rapidly oxidised in air. Te reducing properties of titanous solutions have long been known.

The oxidation potential of the titanous -titanic coupe is 0.1 volt.

$$Ti^3 + H_2O \rightarrow TiO^{2+} + 2H^{2+} + e^-$$

As the H^+ ions are evolved during the oxidation and hence potential of the system is strongly dependent on the acidity of the solution, Kolth off gave the following relationship.

$$E_H = 0.03 + 0.068 \log \frac{\left[Ti^{3+}\right]\left[H^+\right]}{\left[Ti^{4+}\right]} \text{ at } 18^\circ C.$$

Although, titanous sulphate is not oxidised easily as is the chloride, it is more readily hydrolysed and therefore can not be employed in very weakly acid solution.

Both titanous chloride in hydrochloric and medium and sulphate in a sulphuric acid medium have been used as titrants.

A. Titanous Chloride

Take 12 g of titanium metal in a flask and add 800 ml of dilute hydrochloric acid (1 :1 : : v/v). Warm it on a water bath till the metal dissolves. Allow to cool in a stream of hydrogen. Dilute to two litres with freshly boiled and cooled distilled water. The solution is about N/10.

B. Titanous Sulphate

Heat 20 ml of concentrated sulphuric acid and 100 ml of water at 70°C and add 0.6 g of titanium hydride in small amount. When the reaction subsides, boil fur 2 minutes and pour into 900 ml of deaerated water. Filter and store under hydrogen.

STANDARDISATION OF TITANOUS SOLUTION

It is always essential to standardise the solution before use because of its great sensitivity. Titrations must be carried out in the absence of oxygen which is done by bubbling carbondioxide of nitrogen through the titration flask.

First Method

Knectit and Hibbert standardised titanous solution against ferric ammonium sulphate.

To the measured volume of standard ferric ammonium sulphate solution add equal volume of 40 per cent dilute sulphuric acid in the titration flask. Pass a strong stream of carbon dioxide in order to displace the air. Run titanous solution until the yellow colour of the ferric solution has almost disappeared. Now add few drops of ammonium thiocyanate solution and continue the titration until the red ferric thiocyanate disappears. The reaction is carried out at 35°C.

Second Method

Breit examined serval methods for the standardisation and found that direct titration of dichromate gave the most consistent results. Titrate air free standard potassium dichromate solution containing dilute sulphuric acid with titanous solution using 0.5 per cent aqueous solution of sodium diphenyl benzidine sulphonate as indicator.

Applications

Titanous salts are used as reductant in the volumetric estimation of a wide variety of samples.

1. Determination of Ferric Iron

Ferric iron in hydrochloric acid medium may be determined at higher temperature (50 – 60°C) using ammonium thiocyanate as indicator. At the end point the red colour of the thiocyanate disappears.

$$FeCl_3 + TeCl_3 \rightarrow FeCl_2 + TiCl_4$$

2. Determination of Cupric Copper

Cupric copper may be estimated employing thiocyanate and a little of ferrous salts as indicator. The procedure also allows the determination of ferric and cupric ions in presence of each other.

3. Determination of Pentavalent Antimony

Holness and Cornish have recommended the titrations of antimony (V) against titanous sulphate using aqueous solutions of indigo carmine as indicators.

4. Determination of Nitrate

Oldhin described methods for the aliphatic nitrate using methylene blue as indicator.

5. Determination of Perchlorate

Perchlorate may be reduced to chloride by boiling for a long time with titanous solution in a strong hydrochloric acid medium.

Many other oxidising agents such a dichromate, ferricyanide, etc. have been determined against titanous solutions.

6. Nitro Compounds

Nitro, nitroso and azo groups in aromatic compounds may be determined by boiling the respective compounds in hydrochloric acid solution with an excess of titanous chloride and back titrating with ferric solution.

$$RNO_2 + 6TiCl_3 + 6HCl \rightarrow RNH_2 + 6TiCl_4 + 2H_2O$$

$$RNO + 4TiCl_3 + 4HCl \rightarrow RNH_2 + 4TiCl_4 + H_2O$$

$$RN = NR + 4TiCl_3 + 4HCl \rightarrow RNH_2 + RNH_2 + 4TiCl_4$$

But the above method does not give exact result because of the reduction of hydrogen ions. Kolthoff *et al.* have recommended the use of sodium citrate in order to lower the acidity.

7. Organic Peroxides

Yule and Wilson have reported the determination of peroxides by reducing them with ferrous thiocyanate in aqueous acetone and by titrating the ferric ion formed against titanous solution.

8. Sulphoxides

Bernard *et al.* reported that reducing with titanous chloride is most satisfactory for the determination of sulphoxides.

9. Quinone

Quinone may be easily determined by reducing in acid solution to hydroquinone employing methylene blue as indicator.

$$C_6H_4O_2 + 2TiCl_3 + 2HCl \rightarrow C_6H_4O_2H_2 + 2TiCl_4$$

CHROMOUS SALTS

Chromous, *i.e.*, chromium (II) salts are most powerful reductants. They are employed in the form of standard solutions in volumetric analysis. The redox potential of the chromous-chromic system is as follows:

$$Cr^{2+} \rightleftarrows Cr^{3+} + e^-$$

Chromous ions (blue coloured) are so easily oxidised by air to chromic ions (green coloured) that the solution must be stored and delivered form burette under inert gas the solution being titrated must be completely freed form dissolved oxygen.

$$4Cr^{2+} + O_2 + 4H^+ \rightarrow 4Cr^{3+} + 2H_2O$$

Due to the strong reducing action of Cr^{2+} ions, even H^+ ions are reduced.

$$2Cr^{2+} + 2H^+ \rightarrow 2Cr^{3+} + H_2\uparrow$$

The oxidation of Cr^{2+} by H^+ is, however, very slow which is catalysed by various substances, particularly platinum and other finely-divided metals.

PREPARATION OF CHROMOUS SULPHATE SOLUTION

The best method for preparing the chromous salt solution is by the reduction of potassium dichromate solution with amalgamated zinc. Dissolve suitable quantity of dry A.R. potassium dichromate in water and acidify the solution with concentrated sulphuric acid. Now add slowly in several portions of 30 per cent hydrogen peroxide.

The dichromate is thus reduced to chromic ions. Heat the solution to boiling until evolution of oxygen stops. Cool to room temperature and dilute. The reduce the chromic solution by means of suitable quantity of amalgamated granulated zinc. Now pass the hydrogen gas through the solution. The reduction is complete when the solution has acquired a uniform sky-blue colour without any thing of green.

The solution is preserved under inert atmosphere.

Standardisation—Ist Method

Titrations must be carried out potentiometrically because of the deep green colour of the chromic ion. The potential change at the end point in this system is very sharp. The normality of the chromous sulphate solution may be evaluated by titration against pure copper sulphate in a solution containing a large amount of chloride ion. The complex tetra chloro cuprate ion is reduced to chloro complex of univalent copper.

$$[CuCl_4]^{2-} + e^- \rightarrow [CuCl_2]^- + 2Cl^-$$

and then to metallic copper.

$$[CuCl_2]^- + e^- \rightarrow Cu + 2Cl^-$$

The potantiometric titration yields clearly both stages of reduction and the end point can be determined with great accuracy.

2nd Method

The solution may be also standardised against dichromate by adding a slight excess of ferrous ammonium sulphate solution to the dichromate solution in N-sulphuric acid and then titrating the ferric iron with chromium solution potentiometrically.

$$Cr^{+2} + Fe^{+3} \rightarrow Cr^{3+} + Fe^{+2}$$

Analyses Involving Chromous Solutions

Systematic studies on the use of chromous solution for the direct potentiometric determination of certain metals were first undertaken by Zintal *et al.*

Someya determined chlorate and nitro compounds by reducing with an excess of chromous chloride and back titrating with a ferric using thiocyanate solution as indicator.

In more recent times there appears to be a great interest in this titrant. Visual determinations of dissolved oxygen in water, chloroform and of carbon tetrachloride, and potentiometric determination of various organic compounds have been reported.

CHLORAMINE–T

The sodium derivative of N-chloro-p toluene-sulphonamide is known as chloramine – T. It is obtained in a high state of purity as a by-product in the manufacture of saccharin, A review including the physical and chemical properties of chloramine - T has been given by Afaasjer (1952).

Chloramine-T acts as an oxidant in alkaline a well as in acid solution: the reaction mechanism is not yet fully explained. In aqueous solution, chloramine-T is considered to hydrolyse according to the following equation.

$$CH_3C_6H_4SO_2NClNa + H_2O \rightleftarrows CH_3C_6H_4SO_2NH_2 + NaIO$$

The active constituent of chloramine-T is the hypochlorite ion (OCl) which results from the hydrolysis of the reagent. Chloramine-T is preferred to hypochlorite because of its relatively high stability in aqueous solution and since the solution is neutral, no chlorate formation occurs. It has never found use as a titrant, however because it does not possess the range of hypochlorite.

The redox potential of chloramine-T in H_2SO_4 is 1.52 V in natural solution 0.90 V.

Standard Solution

The standard solution is prepared by dissolving the pure substance. The aqueous solution if stored in dark bottles does not change their true normality for several months.

The solution is standardised iodometrically or by titration with an arsenite solution using starch as indicator.

Determination Involving Chloramine-T

(i) *Arsenic and antimony* : The oxidation of trivalent arsenic with chloramine-T proceed according to the equation.

$$AsO^{3-}_3 + CH_3C_6H_4SO_2NClNa + H_2O \rightarrow AsO^{3-}_4 + CH_3C_6H_4SO_2NH_2 + NaCl$$

In a sodium hydrogen carbonate solution arsenite can be titrated with chloramine - T after adding potassium iodide using starch as indicator.

The potentiometric and visual titration of arsenic (III) compounds with chloramine-T have been again thoroughly studied in order to set conditions for precise determinations.

(ii) *Tin* : The determination of tin (II) salts with chloramine-T is carried out in HCl solution in an inert atmosphere. The end point can also be determined amperometrically.

(iii) *Iron* : The determination of iron (II) salts with chloramine -T is carried out potentiometrically in 2 – 3 N HCl solution.

(iv) *Hexacyanoferrate (II)* : Poethke et al. (1952) carried out its determination in strongly HCl medium. The oxidation takes 2 – 3 min.

(v) *Hydrazine* : It is oxidised with chloramine-T according to the equation.

$$N_2H_4 + 2CH_3C_6H_4SO_2NClNa \rightarrow N_2 + 2CH_3C_6H_4SO_2NH_2 + 2NaCl$$

The titration is carried out directly in sodium hydrogen carbonate solution in the presence of iodide-starch as indicator.

(vi) *Thiosulphate* : The potentiometric titration of thiosulphate with chloramine-T (oxidation to sulphate) proceeds best at 50–60°C in solution containing boric acid.

(vii) *Iodide :* The oxidation of iodide with chloramine-T has been recently studied extensively.

In addition to above, cyanide, sulphite, thiocyanate, thallium (I), hypophosphorous acid, hydrogen peroxide, bromate (V), etc. have been also determined.

(viii) *Aldehydes :* Some aldehydes are oxidised quantitatively with chloramine-T in alkaline solution.

$$RCHO + CH_3C_6H_4SO_2NClNa + NaOH \rightarrow RCOONa + CH_3C_6H_4SO_2NH_2 + NaCl$$

(ix) *Uric Acid :* Uric can be titrated with chloramine-T in $NaHCO_3$ solution in the presence of KI, using a-capthoflavone as indicator.

A number of organic compounds like urea, thiourea, oxalic acid, glycerol, glucose, phenols, etc. have been titrated against chloramine–T.

The published work relates with the catalytic action of osmium tetraoxide on the oxidation of chloramine – T in alkaline solution.

MIXED METHODS

(a) *Electrolytic Reduction using metal electrodes :* number of organometallic compounds can be prepared by this method. The various examples are as follows :

(i) By electrolysing triethyl aluminium sodium fluoride with lead cathode.

$$3Pb + 2(Et_3Al)_2NaF \rightarrow 4Al + 3PbEt_4 + 2NaF$$

(ii) Formation of $Sn(CH_2CH_2CN)_4$ if an aq. alkaline solution of acrylonitrile is electrolysed with a tin cathode :

(iii) Formation of $(Me_2CH)_4Pb$ or $(Me_2CH)_4Hg$ if the solution of acetone is electrolysed with lead or mercury cathodes.

(b) *Reaction of diazo alkanes with hydrides and halides :* The following examples illustrate the process involved.

$$CH_2N_2 + HgCl_2 \rightarrow \underset{\text{Chloro methyl chloride}}{ClCH_2HgCl} |+ N_2 \uparrow$$

$$ClCH_2HgCl + CH_2N_2 \rightarrow \underset{\text{Bis (chloro methyl) mercury}}{(CH_2Cl)_2Hg} + N_2 \uparrow$$

(iii) *Decarboxylation* : The general reaction is follows :

$$RCOOM \rightarrow RM + CO_2°$$

The method is applicable to only a few isolated instances. R must be strongly electron withdrawing group, with arsenic and mercury carboxylates the reaction takes place as :

$$Me_2AsOCOCF_3 \rightarrow Me_2AsCF_3 + CO_2 \uparrow$$

$$(CF_3COO)_2Hg \xrightarrow{-CO_2} CF_3COOHgCF_3 \xrightarrow{-CO_2} CF_3HgCF_3$$

PROPERTIES OF ORGANOMETALLIC COMPOUNDS

1. Physical Characteristics

the primary covalent character of most organometallic compounds can be seen in their physical properties. They are generally low-melting solids, liquids or gases at ordinary temperatures. They are usually soluble in non-polar or weakly polar organic solvents such as alkanes, toluene, ether or even alcohol.

2. Chemical Characteristics

The chemical properties (especially thermal stability) vary widely. For example, Me_4Ti decomposes at room temperature while Me_4Si remains unchanged even after many days at 500°C . Reaction rates of these compound also vary considerably, *e.g.,* dimethyl mercury is not attacked at room temperature by oxygen in the air while dimethyl zinc is spontaneously inflammable.

Some of the important chemical characteristics are described as follows :

(a) Stability Considerations

Thermochemical measurements reveal the following order of stability for organometalic compounds having various organic groups attached to a metal :

$$Et - M < M - < Ph - M < CF_3 - M$$

This order has been found to be true for most metals. Higher primary alkyls have been found to be almost of the same order of stability as ethyl but secondary and tertiary alkyls have been found to be less stable. This is partly attributed to the steric factors which do not allow association and hence stabilisation of the molecules by intermolecular secondary bonding forces.

The greater stability of aryls as compared to alkyls has been probably ascribed to those metals with d-orbitals either filled and capable of back donating electrons into the antibonding (π^*) orbital of the aromatic nucleus or empty and hence able to overlap and accept electrons from the aromatic π-orbitals. Such bonding will tend to increase the bond order and hence bond strength.

(b) Kinetic Stability

It is possible to determine the thermodynamic stability of a compound from its standard free energy of formation ΔG^o. But standard entropy data and hence ΔG^o, for organometallic compounds are not available. Therefore, we can use the values of standard enthalpy of formation to indicate approximately the thermodynamic stability of organometallic compounds.

From enthalpy data of organometallic compounds, it follows that most of the lighter elements are thermodynamically stable at room temperature with respect to the decomposition into the elements while those of the heavier metals are unstable to such decomposition.

Most transition metal compounds have inherently more kinetic activity than their non-transition metal counterparts due to the availability of empty lower-energy inner d-orbitals on the metal. For example, tetramethyltitanium undergoes spontaneous decomposition at room temperature while tetramethylsilane is stable upto 500°C.

(c) Stability to Oxidation

All organometallic compounds are thermodynamically unstable to oxidation. Many are also kinetically unstable to oxidation at room temperature.

Kinetic instability to oxidation is attributed to the stability of either empty low lying orbitals, *e.g.*, 4p in Me_3Ga or of a lone pair of electrons as in Me_3Bi. In contrast, the group IVB alkyls have either of these features and behave as separated compounds.

As organometallic derivatives of most of the metals have been destroyed by atmospheric oxygen, they should be protected from its action with great care. As soon as the vapour of trimethyl boron comes in contact with atmospheric oxygen, it catches fire spontaneously and burns to produce diboron trioxide with green edge flame. Controlled oxidation results in the formation of esters of alkyl boric acids. For example, trimethyl boron at pressure below the explosion limit on oxidation produces Me_2BOOMe.

$$Me_3B + O \rightarrow Me_2BOOMe$$

With alkaline hydrogen peroxide, oxidation of organoboron compounds occurs rupturing the B – C bonds resulting in the quantitative formation of an alcohol and boric acid.

If the oxidation reaction is rapid as it has been found with some volatile alkyls of boron and aluminium, there occurs great danger of explosion. But on the other hand volatile alkyls of mercury, silicon, germanium and lead even on prolonged exposure have not been attacked by atmospheric oxygen.

It not essential that if the metals are active towards the atmospheric oxygen, then their organometallic derivatives may also be reactive and vice-versa there are metals which are normally very reactive, form organometallic derivatives, which are inert. For example, ferrocene is not affected by atmospheric oxygen or water though iron in attacked by them. On the other hand, there are groups which are so inert and when bonded to metalloids like silicon make them reactive.

(d) Stablility to Hydrolysis

Organo-derivatives of group IVB and VB have been found to be kinetically stable to attack by water while those of the Groups IA and IIA element and of zinc, cadmium, Al, Ga and In undergo hydrolysis readily. These hydrolysis reactions usually involve the nucleophilic attack by water and thus, get facilitated by the presence of empty low-lying orbitals on the metal atom (*e.g.*, Al) which attract the electron pair of the nucleophile.

$$C_4H_9Li + H - OH \rightarrow C_4H_{10} + LiOH$$

$$CH_3MgBr + H - OH \rightarrow CH_4 + Mg(OH)Br$$

In the tetra-alkyls of Group IVB and trialklys of Group VB elements, metals are having filled low-energy orbitals and nucleophilic attack is not favored. The rate of hydrolysis has been found to depend on the polarity of metal-carbon bond and where this is high (*e.g.*, Me_3Al) rapid hydrolysis takes place. Me_3B is not affected by water at room temperature inspite of the empty 2p-orbitals on the boron atom.

Expansion of coordination number beyond 4 in organometallic compounds of silicon and its congeners Ge, Sn and Pb ordinarily takes place when highly electronegative groups such as halogens are attached to the metal atom. When such electronegative groups are not present as in tetramethylsilane, there exists generally no tendency of the metal atom to expand its coordination number beyond 4 and consequently, tetramethylsilane has been kinetically inert to hydrolysis, oxidation and thermal decomposition.

The majority of the organo derivatives of transition metals have been found to be inert to hydrolysis.

(e) Reaction with Carbon Dioxide

Grignard reagents react with excess of carbon dioxide to yield carboxylic acids. The alkyls of beryllium and aluminium have been found to react similarly with an excess of carbon dioxide to form carboxylic acids. On the other hand alkyls of zinc and cadmium do not react with carbon dioxide.

In general alkyls of all metals having electronegativity below 1.5 will react with carbon dioxide and those having values above 1.5 will not react.

REACTION WITH NON-METALLIC ELEMENTS AND WITH THEIR INORGANIC COMPOUNDS

All reactive organometallic compounds get decomposed by halogen elements to form hydrocarbon halides.

$$C_2H_5Li + Br_2 \rightarrow C_2H_5Br + LiBr$$

$$OHC_2H_4HgI + I_2 \xrightarrow{KI} OHC_2H_4I + HgI_2$$

The lower trialkyl boranes burn in chlorine or bromine but tripropyl borane reacts readily at 150°C to form iodo dipropyl borane.

$$Pr_3^n B + I_2 \rightarrow Pr_2^n BI + Pr^n I$$

Trimethyl borane reacts with chlorine in the gas phase of – 95°C to yield chloromethyl di-methyl borane.

$$Me_3B + Cl_2 \xrightarrow{90^oC} ClCH_2BMr_2 + HCl$$

Under suitable conditions the methyl groups of the methyl chloro silanes may get chlorinated without rupturing the C – Si bonds.

$$CH_3 - Si - + Cl_2 \rightarrow CH_2Cl - Si - + HCl \text{ etc.}$$

Some organic derivative of metalloids react similarly.

Phenyl groups have been slowly cleaved by dry HCl at room temperature or by hot concentrated solutions of it to form benzene.

$$C_6H_5 - Si - + HCl \rightarrow C_6H_5 + Cl - Si$$

If hydrogen chloride is allowed to pass into tributyl borane at 110°C dibutyl chloro borane is formed in quantitative yields.

$$Bu_3^nB + HCl \xrightarrow[AlCl_3]{} Bu_2^n.BCl + C_4H_{10}$$

$$Bu_2^nBCl + HCl \xrightarrow[110^\circ C]{} Bu^n.BCl_2 + C_4H_{10}$$

Sulphur and its compounds react with organic derivatives of active metals such as magnesium to form mercaptans. The oxide of nitrogen oxidizes Grignard reagents more vigorously than sulphur.

$$2C_2H_5MgI + NO_2 \rightarrow Mg - \text{Complex} \xrightarrow{H_2O} \underset{\text{Di - ethyl hydroxyla mine}}{(C_2H_5)_2NOH}$$

Nitrosyl chloride : reacts with phenyl magnesium bromide to produce a poor yield of nitrobenzene.

$$NOCl + C_6H_5MgBr + O \rightarrow C_6H_5NO_2 + Mg(Cl)Br$$

Ammonia reacts with Grignard's reagents to form a hydrocarbon.

$$R - MgX + NH_3 \rightarrow RH + Mg(NH_3)X$$

$$\downarrow 2HCl$$

$$MgClX + NH_4Cl$$

Phosphine also reacts with Grignard's reagent to form a hydrocarbon.

$$2C_2H_5MgBr + PH_3 \rightarrow PH(MgBr)_2 + 2C_2H_6$$

$$\downarrow 3HCl$$

$$\underset{\text{Phosphomium chloride}}{PH4Cl} + 2mgBrCl$$

(g) Reaction with Ethers

Some organometallic compounds have been found to be reactive enough to destroy ethers. Methyl and ethyl ethers react with alkyls of sodium, potassium and cesium as

$$R' - O - R'' + R - Na \rightarrow R' - O.Na + R - R'$$

$$\xrightarrow{H\,\downarrow\,OH}$$

$$R' - OH + NaOH$$

$$CH_3 - O - CH_3 + C_2H_5Na \rightarrow CH_3 - O - Na + CH_2 - C_2H_5$$

$$\xrightarrow{H\,\downarrow\,OH}$$

$$CH_3OH + NaOH$$

In general alkyls of metal with electronegativities less than 0.9 have been found to react with ethers. The products are an alkoxide and hydrocarbon.

(h) Transfer of organic radicals from one metal to another :

Examples involving such reactions are as follows :

$$(C_5H_5)_2Hg + 2Na \rightarrow 2C_5H_5Na + Hg$$

$$3(CH_3)_2Hg + 2In \rightarrow 2(CH_3)3In + 3Hg$$

$$Bu_2Hg + 2Na \rightarrow 2BuNa + Hg$$

$$Ph_2Hg + Be \rightarrow Ph_2Be + Hg$$

In each case there occurs a replacement of a relatively inactive metal from its compound by a more active or electropositive one.

(i) Transfer of Organic Radicals from a Metal to Carbon

Organometallic compounds particularly Grignard's reagents and organolithium compounds find uses in the synthesis of various organic compounds to attack hydrocarbon groups. Thus it is possible to prepare straight chain hydrocarbons form primary alcohol and branched chain hydrocarbons from secondary or tertiary alcohol.

$$CH_3MgBr + R - OH \rightarrow R - CH_3 + Mg(OH)Br$$

$$CH_3MgBr + R - \underset{\displaystyle OH}{\underset{|}{CH}} - R' \rightarrow R - \underset{\displaystyle CH_3}{\underset{|}{CH}} - R' + Mg(OH)Br$$

$$CH_3MgBr + R - \overset{\displaystyle R'}{\overset{|}{\underset{\displaystyle OH}{\underset{|}{C}}}} - R'' \rightarrow R - \overset{\displaystyle R'}{\overset{|}{\underset{\displaystyle OH}{\underset{|}{C}}}} - R'' + Mg(OH)Br$$

As mangesium has a great affinity for oxygen and the linkage between Mg and C is rather loose and so it can be easily ruptured, the primary alcohols can be prepared from formaldehyde in the following way :

ORGANOMETALLIC COMPOUNDS CLASSIFICATION

The organometallic compounds may be classified into following classes

1. Olefin Complexes of Metals

In 1830 Zeise was successful in preparing first organometallic compound of platinum *i.e.*, $K[PtCl_3.C_2H_4]H_2O$. It has $[PtCl_3.C_2H_4]^-$ ion. Later on it was found that certain other metal halides or ions particularly those of Cu^+, Ag^+, Hg^{2+}, Pd^{2+} etc. also formed such complexes when treated with a variety of olefins. For example,

$$\underset{\text{(Butadiene)}}{C_4H_6} + Fe(CO)_5 \xrightarrow{80^\circ C} C_4H_6Fe(CO)_3 + 2CO\uparrow$$

$$\underset{\text{(Cycloheptatriene)}}{C_7H_8} + Mo(CO)_6 \xrightarrow{\text{reflux}} C_7H_8Mo(CO)_3 + 3CO\uparrow$$

2. Acetylene Derived Complexes

With the analogy of binding of olefins, it might be expected that acetylenes (RC ≡ CR (There are two π-bonds at right angle to each other) would be bound with a metal atom *e.g.*, cobalt carbonyl reacts with various types of acetylenes.

$$RC \equiv CR' + Co_2(CO)_8 \rightarrow Co_2(CO)_6.(RC \equiv R') + 2CO\uparrow$$

The diphenyl acetylene derivatives have been studied by X-ray diffraction method and it is found that two Co atoms get bonded with two carbon atoms of acetylene and the angle between the Co atoms and C–C axis, if the acaetylene, is about 90°.

3. Cyclopentadiene Complexes of Metals

These compounds have been discussed separately.

4. Complexes of Other Symmetrically Delocalised Hydro-carbons

Cyclopentadiene ion (C_5H_5) has a group of 5-π electrons, other ring systems are expected to form sandwich type compounds similar to cyclopentadiene compounds.

All of these systems (*e.g.*, cyclobutadineyl, arene, cycloheptatrienyl etc. have been found to bond tow metal atoms although $C_2H_5^-$ and C_6H_6 give L_2M type of compounds.

(a) Cyclobutadienyl Metal Complexes

Methyl substituted nickel compound was the first compound of this type.

$$2(CH_3C)_4Cl_2 + 2Ni(CO)_4 \rightarrow [(CH_3C)4NiCl_2]_2 + 8CO \uparrow$$

(b) Arene Metal Complexes

Fisher was able to prepare many new arene metal complexes. Dibenzene chromium, $(C_6H_6)_2Cr$ has been prepared from the Grignard reaction of $CrCl_3$. In general an aromatic hydrocarbon is refluxed with a transition metal halide or carbonyl in the presence of Al powder (acts as a reducing agent and halogen acceptor).

$$2C_6H_6 + CrCl_3 \xrightarrow{Al} (C_6H_6)_2Cr + AlCl_3$$

$$C_6H_6 + Mn(CO)_5Cl + AlCl_3 \rightarrow C_6H_6Mn.(CO)_3^+.AlCl^-_4$$

$$C_6H_5CH_3 + Mo(CO)_6 \rightarrow C_6H_5CH_3Mo(CO)_3 + 3CO \uparrow$$

$(C_6H_6)_2Cr$ is unstable to air. X-ray, electron diffraction study on this compound suggest that its structure is as of ferrocene (*i.e.,* sandwich structure).

(c) Cycloheptatrienyl Metal Complexes

Cycloheptatrienyl ($C_7H^+_7$) metal complexes could not be prepared by the direct reaction on $C_7H^+_7$ and the metal salt. Such compounds may be prepared indirectly.

For example,

$$\pi C_5H_5.V.(CO)_4 + C_7H_8 \rightarrow \pi C_5H_5.\ VpC_7H_7$$

$C_7H_8.Mo(CO)_3 + (C_6H_5)_3C^+BF^-_4 \rightarrow \pi C_7H_7Mo(CO)^+_3BF^-_4 +$ $(C_6H_5)_3CH$

(d) Heterocyclic Sandwiched Complexes

The heterocyclic rings which have a significant aromatic character also form sandwiched type bonds with metal atoms. For example, thiophene and pyrrole compounds $C_4H_4S.Cr(CO)_3$ and $C_4H_4NMn(CO)_3$ are analogous to $C_6H_6Cr(CO)_3$ and π- $C_5H_5Mn(CO)_3$ respectively. Such type of compounds could be prepared by indirect method.

$$\pi - C_5H_5Fe(CO)_2I + C_4H_4NK \xrightarrow{C_6H_6} \pi.C_5H_5Fe.\pi.C_4H_4N$$

$$FeCl_2 + C_5H_5Na + C_4H_4NNa \xrightarrow{THF} \pi.C_5H_5Fe.\pi.C_4H_4N$$

CYCLO PENTADIENYL COMPLEXES

These form another important class of organometallic compounds of transitions of transition metals which involve C_5H_5 group. In the year 1948 S.A. Miller and his colleagues reported that cyclopentadiene reacted with the reduced iron catalyst to give stable orange crystal which is now known as ferrocene. They did not publish anything about the compound. T.J. Kealy and P. L. Pension came across this extremely stable compound in 1961 while attempting to prepare fulvalene $C_{10}H_8$ be $FeCl_3$ catalysed decomposition of C_5H_5MgI. The cyclopentadienyl radical gives three distinct classes of compounds.

(a) Ionic Complexes

Cyclopentadiene is a weak acid. It therefore undergoes combination with a number of bases giving salts of the cyclopentadienide ion C_5H_5. They are also known as cyclopentadienides and are formed with the metals of groups IA, IIA, IIIA and a part of group IVA, $C_5H_5Na^+$.

(b) p-Cyclopentadienyl Complexes

Cyclopentadienyl ion like other ring systems which give sandwich compounds, is having the aromatic sextet of π electrons. Therefore it gives rise to complexes which have the cyclopentadienyl ring covalently bonded to metal. The metal is lying below the plane of the carbon atoms, and is usually equidistant from the five equivalent carbon atoms. These complexes are formed with atoms throughout the transition metal series *e.g.*, $(C_5H_5)_2Fe$.

(c) σ-Cyclopentadienyl Complexes

Besides these the cyclopentadienyl radical is also able to bind to the metal by one carbon atom by a covalent 2-electron bond thus giving rise to σ cyclo-dienyl complexes. These complexes are formed by the metals like iron when all but one coordination position is blocked *e.g.*, Fe(σ – Cp) $(CO)_2$(σ – Cp).

There occurs no abrupt transition between the ionic and covalent metal bond. In many π-cyclopentadienyl it has been observed that there exists a considerable ionic character. Even ferrocene has a charge separation between the metal atom and the πC_5H_5 rings.

In this chapter we will mainly describe the organometallic compounds of transition metals.

(d) π-Cyclo Pentadienyltransition Metal Complexes

All the element of the three transition series except hafnium form cyclo pentadienyl compounds of the general formula $[(\pi - C_5H_5)_2M] X_{n-2}$ where n stands for the oxidation number of the M and X is a univalent anion.

Except manganese the elements of the first transition series give rise to neutral compounds of the general formula $(\pi - C_5H_5)_2M$ as well as cationic species. In the second and third transition series which involve the +2 oxidation state of the metal neutral compounds have been obtained only for Ru and Os. In the +3 oxidation state of the metal cationic species as $(\pi - C_5H_5)_2Co^+$ are obtained. In the +4 and +5 oxidation state of the metal di-π-cyclopentadienyl metal halides of titanium and vanadium groups have also been reported.

System of Notation

In 1968 Cotton suggested a system of notation for organometallic compounds. In this system the number of carbon atoms that are attached to a metal atom is specified by a prefix such as trihapto -, tetrahpto -, pentahapto-etc. These prefixes can be represented as h^{-3}, h^{-4}, h^{-5} etc. It necessary, the bound carbon atoms may by specified by number, using a conventional naming and numbering scheme for the organic group attached to the metal atom.

For example, a commonly known compound ferrocene $(C_5H_5)_2$ Fe is bis (pentaheptocyclopentadienyl) ion, $(h^5 - C_5H_5)_2Fe$ because iron atom is attached (linked) with all the five carbon atoms of $C_5H_5^-$. Similarly titanium derivative of cyclopentadienyl $(h^5 - C_5H_5)_2$ $(h^1 - C_5H_5)_2Ti$ is named as bis (pentaheptocyclo pentadienyl) bis (monohepto cyclopentadienyl) titanium (IV) because out of four $C_5H_5^-$ ions, titanium is linked with all the five carbon atoms of two $C_5H_5^-$ ions.

Methods of Preparation

(1) Reaction of alkali metal cyclo pentadienides with metal complexes: A solution of the alkali metal cyclo pentadienide is prepared by the action of dispersed alkali metal on cyclo pentadinene in tetrahydro furan.

$$2C_5H_5 + 2Na \rightarrow 2C_5H_5^- Na^+ + H_2$$

Now this solution is used for the preparation of number of organometalic compounds of transition metals as :

(a) By reacting freshly prepared solution of sodium or potassium cyclo pentadienide in tetrahydrofuran or in a poly-ether like ethylene glycol dimethyl ether and anhydrous metal halides.

$$2C_5H_5^- Na^+ + FeCl_2 \rightarrow (C_5H_5)_2Fe + 2NaCl$$

$$4C_5H_5^- K^+ + UCl_4 \rightarrow (C_5H_5)_4U + 2KCl$$

The product after the removal of solvent ether is extracted by solvent extraction method or by direction sublimation from the residue in vacuum in the case of air sensitive compounds.

(b) By reacting metal carbonyls with alkali metal cyclo pentadienides in tetrahydro furan when π-cyclo pentadienyl metal carbonyl complexes are obtained.

$$W(CO)_6 + C_5H_5Na^+ \xrightarrow{\text{T.H.F.}} \left[\pi.C_5H_5W(CO)_3\right]^- Na^+$$

π-Cyclo pentadienyl metal carbonyl complexes are also prepared by reacting metal halides with sodium cyclo pentadienide in the presence of carbon monoxide under pressure.

$$TiCl_4 + C_5H_5^-Na^+ \rightarrow (\pi - C_5H_5)_2Ti(CO)_2$$

(2) *Reaction between cyclo pentadiene and metal complexes :* (a) The di π-Cyclo pentadienyl complexes of iron, cobalt, nickel, titanium etc. are also obtained by the reaction of respective anhydrous halides with cyclo pentadiene in the presence of a bases like diethyl amine, piperidine or pyridine which serves as the solvent as well as the halogen acceptor.

$$FeCl_2 + 2C_5H_6 + (C_2H_5)_2NH \xrightarrow{\text{In excess}}$$

$$(\pi - C_5H_5)_2 Fe + 2(C_2H_5)_2 NH_2Cl$$

$$NiCl_2 + 2C_5H_6 + 2(C_2H_5)_2 NH \xrightarrow{\text{amine}}$$

$$(\pi - C_5H_5)_2 Ni + 2(C_2H_5)_2 NH_2Cl$$

(b) π-Cyclo penta dienyl metal carbonyl complexes are also obtained by the reaction of cyclo pentadiene with metal carbonyls under suitable conditions. For example when $Fe(CO)_5$ is refluxed with cyclo pentadiene, $[\pi - C_5H_5Fe(CO)_2]_2$ is obtained in good yields. Under carefully controlled conditions the hexa carbonyls of molybdenum and tungsten give rise to the binuclear complexes of composition $[\pi\text{–}C_5H_5M(CO)_3]_2$.

(3) *Reaction between fulvenes and metal carbonyls :* Fulvenes when reacted with the carbonyls of the VIth group metals yield substituted π-Cyclo pentadienyl complexes as :

A similar reaction occurs between dialkyl fulvenes and dicobalt octacarbonyl.

Properties

Important properties of some cyclopentadienyl complexes of the first row transition elements are given in Table 5.2.

Table 5.2 : Some properties of cyclopentadienyl complexes of the first row transition elements.

Element	*Compound*	*U.P.(0°C)*	*Colour*	*No. of unpaired electrons*	*Magnetic moment (B.M)*
Ni^{II}	$[(C_5H_5)_2Ni]$	173	Green	2	2.86
Ni^{III}	$[(C_5H_5)_2Ni]^+$	–	Yellow	1	1.75
Co^{II}	$[(C_5H_5)_2Co]$	173	Purple	1	1.76
Co^{III}	$[(C_5H_5)_2Co]^+$	–	Yellow	0	Diamagnetic
Fe^{II}	$(C_5H_5)_2Fe$	174	Orange	0	Diamagnetic
Fe^{III}	$[(C_5H_5)_2Fe]^+$	–	Blue	1	2.26
Mn^{II}	$(C_5H_5)_2Mn$	173	pink	5	5.9
Cr^{II}	$(C_5H_5)_2Cr$	173	Scarlet	2	2.84
Cr^{III}	$[(C_5H_5)_2Cr]^+$	–	Green	3	3.81
V^{II}	$(C_5H_5)_2V$	168	Purple	3	3.86
V^{III}	$[(C_5H_5)_2V]^+$	–	Purple	2	2.86
Ti^{II}	$(C_5H_5)_2Ti$	130	Green	0	Diamagnetic
Ti^{III}	$[(C_5H_5)_2Ti]^+$	–	Green	1	1

All are crystalline solids which are generally coloured *e.g.*, $(\pi\text{–}C_5H_5)_2Ni$ is bright green $(\pi\text{–}C_5H_5)_2TiCl$ has bright red crystals $(\pi\text{–}C_5H_5)_2Fe$ has orange crystals. The neutral compounds of Ti, Cr, Fe, Co, Ni are soluble in hydrocarbon solvents. They readily sublime in vaccum. Most of the sandwich complexes are thermally stable and melt without decomposition at about 173°C. The titanium compounds is thermally unstable and decomposes at about 140°C before melting.

The stability of the cyclopenta dienyls towards oxygen has been found to vary widely. Ferrocene is stable toward molecular oxygen at room temperature while π-cyclo penta dienyl metal oxide derivatives are found if these complexes are put to controlled oxidation.

Practically all the compounds reacts in acid solution with a variety of oxidising agents to yield cations.

$$(\pi\text{–}C_5H_5)_2M \rightarrow (\pi C_5H_5)_2M^+ + e^-$$

These cations can be isolated by precipitation with suitable anions.

The V^{2+}, Fe^{2+}, Ni^{2+}, Co^{2+} compounds are not affected by water. However, they are merely oxidised to $M(C_5H_5)^+_2$ by 3M – HNO_3. The hydrogen atoms of π-cyclo penta-dienyl ligands are normally not replaceable and do not exchange with D_2O. However at higher temperatures $Ca(OD)_2$ reacts with $(\pi\text{–}C_5H_5)_2Fe$ to form deutero ferrocenes.

$$(\pi - C_5H_5)_2\,Fe + 5VCa(OD)_2 \rightarrow \underset{\text{Deutero ferrocene}}{(C_5H_5)_2\,Fe} \qquad +5Ca(OH)_2$$

The di π-cyclo pentadienyl complexes have been found to be even resistant to catalytic hydrogenation. Ferrocene is even more resistant to hydrogenation then benzene. However it gets reduced by the lithium in ethyl amine with the rupturing of linkage. The neutral compounds of V, Mn, Co and Ni on treatment with CO or NO under pressure lose one ring to form carbonyl or nitrosyl complexes.

$$(\pi - C_5H_5)_2\,Co + CO \xrightarrow[50-200^\circ C]{100-200 \text{ atm. pressure}} (\pi C_5H_5)Co(CO)_2$$

$$(\pi - C_5H_5)_2\,Ni + NO \xrightarrow[\text{hexane}]{1 \text{ atm. pressure}} \pi - C_5H_5Ni(NO)$$

The cyclo pentadienyl rings in ferrocene and related compounds have an aromatic character. This is evident from the measured C – C bond distances in the complexed cyclo-pentadienyl ring and the replacement of the hydrogen atoms in it by acyl and sulphonic groups. Condensation reactions with aldehydes and reaction with diazonium compounds also support the view that complexed cyclo-pentadienyl rings have aromatic character. The aromatic character of complexed cyclo pentadienyl rings strongly suggests that each ring has a delocalised sextet of electrons just as in benzene. The C_5H_5 group has five localised electrons and, by gaining one electron from the metal, is converted to C_5H_5 possessing six non-localised electrons.

The cations $(\pi\text{–}C_5H_5)_2Co^+$, $(\pi\text{–}C_5H_5)_2Ir^+$,$(\pi\text{–}C_5H_5)_2Rh^+$ having metal in the +3 oxidation state have been reported to be stable in aqueous solution. The $(\pi\text{–}C_5H_5)_2Fe^+$ ion is moderately stable in acid solutions and gets hydrolysed giving ferrocene. The $(\pi\text{–}C_5H_5)_2Co^+$ ion has been very stable and is unaffected by concentrated nitric acid even on heating. The ions $(\pi\text{–}C_5H_5)_2V^+$ and $(\pi\text{–}C_5H_5)_2Ti^+$ get oxidised by dilute acids, the oxidation occurs. More rapidly in air. The $(\pi\text{–}C_5H_5)_2Ni^+$ is thermally unstable in solution and gets decomposed slowly by air and readily by base. All organometallic cations $[(\pi\text{–}C_5H_5)_2M]^+$ yield precipitates with silico tungstic acid in aqueous solution. With concentrated solutions precipitates may also be obtained with anions like $(PtCl_6)^-$, O^-_3, picrate, and Reineckate.

When $(\pi\text{–}C_5H_5)_2Co^+$, is treated with $LiAIH_4$, it gets reduced to give cobaltocene, With phenyl-lithium it reacts in the following manner. The di and trihalides of di π-cyclo pentadienyl complexes of Tl^{+4}, Zr^{+4}, Nb^{+5}, have also been reported, When $(\pi\text{–}C_5H_5)_2NbBr_2$ get hydrolysed in the presence of air it gives a well defined orange compound of composition $(\pi\text{–}C_5H_5)_2Nb(OH)Br_2$

The inferred spectra of these compounds have been found to be very similar and reveal a high degree of symmetry. The visible and ultraviolet spectra, on the other hand, are quite dissimilar and indicate very different electronic configurations. The π-cyclo–pentadienyl carbonyls are also highly coloured and soluble inorganic solvents. They follow 18-electron rule : since the cyclopentadienyl ring is less stable than a carbonyl group, it frequently survive the action of agents capable of removing carbonyl groups or replace other ligands.

$$\pi-C_5H_5V(CO)_4 \xrightarrow{Cl_2O_2} \pi-C_5H_5VOCl_2 + 4CO_2$$

$$\pi-C_5H_5Mo(CO)_3Cl + Cl_2 \rightarrow C_5H_5Mo(CO)_2Cl_3 + CO$$

$$\pi\text{–}C_5H_5Mo(CO)_2Cl + NH_3 \rightarrow [\pi - C_5H_5Mo(CO)_3NH_3]^+Cl^+$$

The metal-metal bonds in binuclear cyclopentadienyl carbonyl derivatives may be cleaved by oxidising or reducing agents.

$$[\pi\text{–}C_5H_5Fe(CO)_2]_2 + Cl_2 \rightarrow 2\pi - C_5H_5Fe(CO)_2Cl$$

$$[\pi\text{–}C_5H_5Fe(CO)_2]_2 + Na \rightarrow 2Na^+ [\pi - C_5H_5Fe(CO)_2]^-$$

SALINE OR SALT-LIKE HYDRIDES

These hydrides are formed by elements of very low electronegativities which are sufficiently strong reducing agents to transfer electrons to the hydrogen atoms. The elements of I A, II A and III A form hydrides of this type having composition MHx where x is the group valency of the metal.

These are colourless, crystalline compounds generally insoluble in common solvents. They, however, dissolve in molten halides without reaction. The electrovalent character of these hydrides is evident from their high melting and boiling points, their high electrical conductivity in the fused state and ionic lattices possessed by them. They react with water or dilute acids to evolve hydrogen. These are generally prepared by the combination of metals with hydrogen at 150-800°C.

$$2Li + H_2 \xrightarrow{600°C} 2LiH$$

$$2Na + H_2 \xrightarrow{400°C} 2\,NaH$$

$$Ca + H_2 \xrightarrow{300°C} CaH_2$$

(A) Hydrides of the Alkali Metals (Ist A)

(a) **Lithium Hydride** : It is the most stable of the alkali metal hydrides. It is prepared by the action of hydrogen on molten lithium at 700-800°C.

$$2Li + H_2 \xrightarrow{700-800°C} 2\,LiH$$

Lithium hydride may also be prepared by heating lithium nitride in a current of hydrogen.

$$2Li_3N + 3H_2 \rightarrow 6LiH + N_2$$

Properties

(i) It is a white solid. It reacts with water forming lithium hydroxide and hydrogen.

$$2LiH + 2H_2O \rightarrow 2\,LiOH + H_2$$

(ii) When an ethereal solution of lithium hydride is treated with anhydrous $AlCl_3$, $LiAlH_4$ is formed.

$$4LiH + AlCl_3 \rightarrow LiAlH_4 + 3LiCl$$

(iii) Electrolysis of lithium hydride yields lithium at the cathode and hydrogen at the anode.

(iv) With B_2H_6 in ether medium, it reacts to form lithium borhydride.

$$2LiH + B_2H_6 \rightarrow 2LiBH_4$$

Structure

It is a white crystalline solid with a face-centred cubic lattice (*i.e.*, sodium chloride). The metal-hydrogen bond distance has been determined by X-ray diffraction.

(b) **Sodium Hydride :** Sodium hydride can be best prepared by the action of hydrogen on molten sodium at about 250°C.

$$2Na + H_2 \rightarrow 2NaH$$

It has a face-centred cubic lattice. Sodium hydride is used as a reducing or condensing agent in organic chemistry. For example, benzaldehyde condenses with methyl acetate to give methyl cinnamate in the presence of sodium hydride.

(c) **Potassium Hydride :** Pure potassium hydride may be made by the action of a stream of hydrogen at 400°C on potassium.

$$2K + H_2 \rightarrow 2KH$$

It forms white needles and has the sodium chloride structure.

(d) **Rubidium and Caesium Hydrides :** Both the hydrides can be formed by the direct action of hydrogen on the metal at about 680°C.

$$2Rb + H_2 \rightarrow 2RbH$$

$$2Cs + H_2 \rightarrow 2CsH$$

Both these hydrides possess the sodium chloride structure.

(B) Hydrides of the Alkaline Earth Metals

These hydrides are similar to alkali metal hydrides in the mode of their preparation.

$$Ca + H_2 \xrightarrow{800°C} CaH_2$$

$$Sr + H_2 \xrightarrow{800°C} SrH_2$$

$$Ba + H_2 \xrightarrow{850°C} BaH_2$$

X-ray diffraction shows that calcium ions in calcium hydride are in a slightly distorted hexagonally close-packed arrangement.

The unit cell of strontium hydride is orthorhombic like that of calcium hydride and possesses a similar distorted hexagonally closed packed structure with two kinds of non-equivalent hydride ions. Barium hydride has a similar structure to calcium hydride.

(C) Hydrides of the Third-A Metals

Aluminium hydride lies on the borderline between ionic and covalent hydrides. Aluminium hydride is prepared by treating an ethereal solution of aluminium chloride with lithium hydride or lithium aluminium hydride.

$$AlCl_3 + 3LiH \xrightarrow{\text{Either}} AlH_3 + 3LiCl$$

$$AlCl_3 + 3LiAlH_4 \xrightarrow{\text{Either}} 4AlH_3 + 3LiCl$$

The structure of aluminium hydride is not known but it seems likely that each aluminium is surrounded by six hydrogen atoms in a three-dimensional bridged net work.

MOLECULAR OR VOLATILE OR COVALENT HYDRIDES

These hydrides are formed by the highly electronegative elements (*e.g.*, the non- metals of 3rd B, 4th B, 5th B, 6th B and 7th B groups) by sharing electrons with hydrogen. These covalent hydrides are charaterised by molecular lattices. They possess low melting and boiling points. They are volatile in character and are non-conductors indicating covalent bonding in their molecules. Some of the hydrides in this class contain bridging hydrogens; *e.g.*, the boron hydrides. These are prepared by the following methods:

(i) By direct combination of elements with hydrogen.

$$H_2 + F_2 \rightarrow 2HF$$

(ii) By the action of hydrogen on a metal or its compound.

$$AsCl_3 + 6H \rightarrow AsH_3 + 3HCl$$

$$SbCl_3 + 6H \rightarrow SbH_3 + 3HCl$$

$$GeCl_4 + 8H \rightarrow GeH_4 + 4HCl$$

(iii) By the action of water or dilute acid on boride, carbide, nitride, silicide, etc. of metals.

$$Mg_3B_2 + 6HCl \rightarrow 3MgCl_2 + B_2H_6 \uparrow$$

$$Mg_2Si + 4HCl \rightarrow 2MgCl_2 + SiH_4 \uparrow$$

$$Al_4C_3 + 2H_2O \rightarrow 2Al(OH)_3 + 3CH_4 \uparrow$$

$$Mg_3N_2 + 6H_2O \rightarrow 3Mg(OH)_2 + 2NH_3 \uparrow$$

(iv) By the reduction of the appropriate halide in ether solution with lithium aluminium hydride.

$$SnCl_4 + LiAlH_4 \rightarrow LiCl + AlCl_3 + SnH4$$

$$GeCl_4 + LiAlH_4 \rightarrow LiCl + AlCl_3 + GeH_4$$

A number of covalent hydrides are formed by boron, carbon and silicon.

BORON HYDRIDES

Introduction

None of the group III elements react directly with hydrogen but several interesting hydrides are known. Boron forms a series of volatile compounds with hydrogen. These compounds are known as boron hydrides or more commonly boranes, by analogy with alkanes (hydrcarbons). The formulae of the boranes are $B_n H_{n+4}$ and $B_n H_{n+4}$. Boranes with the general formula $B_n H_{n+4}$ are more stable than the boranes with the general formula $B_n H_{n+6}$. Among the well known boron hydrides are diborane (B_2H_6), tetraborane ($B_{10}H_{14}$, $B_{10}H_{16}$).

Nomenclature of Boron Hydrides

According to the rules of IUPAC system, the number of boron atoms in the molecule is indicated by a greek numerical prefix. The boranes are named as di-, tri-, tetra- penta-, hexa- boranes and so on according to the number of boron atoms followed by the numeral which of indicates the number of hydrogen atoms. For example, B_5H_9 is called penta borane-9. Here penta indicates the number of boron atoms and 9 indicates the number of H-atoms. For example,

Name IUPAC	*Formula*	*M.P. °C.*	*B.P. °C*
Diborane-6	B_2H_6	– 165.6	–92.15
Tetraborane-10	B_4H_{10}	– 120.0	18
Pentaborane-9	B_5H_9	– 46.6	48
Pentaborane-11	B_5H_{11}	– 123.0	63
Hexaborane-10	B_6H_{10}	– 62.3	82
Hexaborane-12	B_6H_{12}	– 82.3	80-90
Decaborane-14	$B_{10}H_14$	– 99.7	213

The polyboranes and their derivatives may he considered to consist of two general classes :

(1) **Closed Structures :** These are the structures with boron skeletons which are polyhedral having all triangular faces. The members of this series are designated by the prefix closo.

(2) **Non-closed Structures :** The members of this series are designated by the prefix nido.

The boranes are also grouped into two classes namely :

(i) B_nH_{n+4}, (n=2, 5, 6, 8, 10 and 18). The boranes of this class are quite stable and have high m.pts. Examples of boranes of this class are diborane, B_2H_6; penta borane- 9, B_5H_9; hexahorane-10, B_6H_{10}; octaborane-12, B_8H_{12}; pentahorane-11, B_5H_{11} etc.,

(ii) B_nH_{n+6} (n=4, 5, 6, 9, and 10). The boranes of this class are unstable and have low melting points. Examples of boranes of this class are tetraborane-10, B_4H_{10}; pentahorane-11, B_5H_{11}; hexahorane-12, B_6H_{12}; enneaborane-15, B_9H_{15}; decahorane-16, $B_{10}H_{16}$ etc.

Preparation of Boron Hydrides

(A) Preparation of a Mixture of Boron Hydrides

(i) **The Original Stock's method :** When Stock treated magnesium boride with dilute aqueous acid, he got B_4H_{10}, B_5H_9, B_6H_{10} and $B_{10}H_{14}$ and other hydrides. The chief component of the volatile mixture obtained by the action of HCl on Mg_3B_2 was B_4H_{10} and traces of B_2H_6.

$$2Mg_3B_2 + 12HCl \rightarrow 6MgC_{12} + B_4H_{10} + H_2$$

These hydrides may he separated by liquefying the gaseous mixture by passing it through a vessel immersed in liquid air and then fractionating the liquid obtained. B_2H_6 is not present in the gaseous mixture obtained by Stock because it is easily decomposed by water.

(ii) **Wurtz's Method :** This method has been used to prepare B_4H_{10} which is obtained by the action of C_2H_5I on sodium amalgam (Na/Hg).

$$2C_2H_5I + 2Na/Hg \rightarrow B_4H_{10} + 2NaI.$$

(iii) **Electric Discharge Method :** When diborane is decomposed in a silent electric discharge in the presence of an inert gas, it gives

B_4H_{10} 40%

B_5H_9 20%

B_5H_{11} 30%

B_9H_{15} Small, quantity.

(iv) **By pyrolysis of diborane :** By this method different hydrides can be prepared. This provides the best route to B_4H_{10}, B_5H_9, B_5H_{11} and $B_{10}H_{14}$. The mechanists of pyrolysis is complex although BH_3 is certainly involved and by varying the temperature, the rate of cooling of the gases, and the concentration of additional components such as H_2 or dimethyl ether, the formation of particular hydrides can be enhanced.

Thus, B_5H_{11} is formed in maximum yield at about 100°C but B_5H_9 is formed more efficiently at 240°C in the presence of H_2. Pyrolysis of a mixture of B_2H_6 and B_6H_{10} gives a small yield of B_6H_{12}, one of the less easily made hydrides.

(v) Another method is by acid-base reaction, *i.e.*, acidification of the salts of $B_3H_8^-$ with phosphoric acid gives small yield of several uncommon boranes including B_8H_{18} and B_6H_{12}. Pentaborane is readily disproportionated by bases; ether gives B_6H_{10} and hexamethylenetetramine gives n-B_9H_{15}.

STRUCTURE

Ferrocen and other related metallocenes have two C_5H_5 rings and their formula, since it involves mainly the π-type overlap, is written as $(\pi - C_5H_5)_2M$. Such compounds have been remarkable for following two reasons.

(i) Sandwich Structure

The metal atom lies symmetrically in between the two C_5H_5 rings. The sandwich structure was first of all postulated by Wilkinson. Rosenblum, Whitting and Woodward and later by Fischer and Pfab.

(ii) Aromaticity

The molecule acts as an aromatic system; for example, it undergoes Friedel-Crafts acetylations. There are many other compounds also, which have sandwich structure, but have only one C_5H_5 rings, like π–cylopentadienyl-chromium (II) dinitrosyl chloride $\pi–C_5H_5Cr(NO)_2Cl$.

However, here we will first discuss the metallocenes having two C_5H_5 rings, with the main emphasis on ferrocene.

Bonding in Metallocenes

The metal atom in ferrocene and other metallocences has been sandwiched between two parallel, plane, regular pentagonal C_5H_5 rings, in which all C – C distances are equal (1.40Å). In each ring, each carbon is in sp hydridized state, the three sp^2 orbitals having been used for bondings with two other carbon atoms and a hydrogen atom. The remaining p orbitals of the five carbon atoms, having an electron each, get involved in a delocalized π – bonding engulfing the whole ring. Thus, the rings contain five π – electrons. In the ferrocene molecule (incorporating both rings), all Fe-C distances have been found to be equal (2.045 $\pm$ 0.01A). represent the common method of depicting the structure of ferrocene. In the figure, the circles with in the rings represent the aromatic character of the rings and the lines from iron atom reveal that metal is bond to the whole rings rather than to any particular carbon atom of the rings.

Any theory postulated for explaining the bonding in metallocene molecules must be capable of explaining the above mentioned characteristics.

Before evolving any theory of bonding in such compounds, it becomes necessary to consider the symmetry properties of the ring π–orbitals and the metal orbitals. These properties are important because only the orbitals of the same symmetry can under go combination to form molecular orbitals. It may be said here that the two rings lie in the xy plane and the line passing through the metal atom and the centres of the two parallel rings, fall on axis. The symmetry properties of the metal and ring π–orbitals first gave by Dunitz and Orgel. They are as follows :

Ring Orbitals

The five Pz orbitals of each ring, each having one electron and extending above and below the plane of the ring (as in case of benzene) may undergo combination in five ways to from five delocalized molecular orbitals. The most bonding of them belongs of A_1 symmetry, followed by a double degenerate set of orbitals of E_1 symmetry and the most energetic molecular orbitals are again a double degenerate set of E_2 symmetry. A_1 orbital does not have nodal plane perpendicular to the xy plane of the ring, E_1 have one such nodal plane and E_2 have two. The respective ring orbitals are shown in Fig. 5.16.

It is obvious that the energy of these orbital increases in the order $A_1 < E_1 < E_2$.

In the metallocenes, the metal atoms interact with tow rings, simultaneously. Therefore it is better to find out the relative orientations of the various orbitals of the same symmetry of the two rings with respect to the atom, which forms the centre of symmetry of the molecule. This can be carried out by linearly combining the orbitals of the two rings. The linear combination of the π–orbitals of the rings. From the examination of the symmetry properties of these composite orbitals it would reveal which orbitals of the ring system could combine with the atomic orbitals of the metal.

The composite orbitals formed are A_{1g} and A_{1u} (non-degenerate); E_{1g} and E_{1u} (each double degenerate); E_{1g} and E_{1u} (each double degenerate) and E_{1g} and E_{1u} (each double degenerate). These composite molecular orbitals. As the A_1 and E_1 grade and ungrade orbital have been of lower energy, the electronic configuration of the ring π – orbitals can be put as follows . $(A_{1g})^2 (A_{1u})^2 (E_{1u})^4 (E_{1g})^2 (E_{2g})^0 (E_{2g})^0$

Metal Orbitals

The metal orbitals in question are-nd, (n + l) s and (N + 1) p. They have been classified into following symmetry types.

d^2_z	-	A_{1g}
d_{xz}, d_{yz}	-	E_{1g}
$d_{xy}, d_x{}^2 - {}_y{}^2$	-	E_{2g}

Moffitt and later Lichr and Ballhausen postulated that the two orbitals of A_{1g} symmetry (d^2_z and s) can undergo mixing to give two hybrid orbitals, $(DS)_x$ and $(DS)_y$. It is evident from Fig. 5.5 that the molecular axis being the z axis, $(DS)_x$ is lying in the region of high electron density due to C_5H_5 rings, whereas $(DS)_y$ is largely removed from this region.

Therefore, the metal-electron density would tend to shift to $(DS)_y$ from $(DS)_z$. Under these circumstances, the ground state configuration of a metal atom,. say iron, would become as follows :

$$(DS)_y{}^2 (3d^2{}_x - y^2)^2 (3d_{xy})^2 (3d_{xz})^1 (3d_{yz})^1 (DS)^0{}_x (4P_x)^0 (5P_x)^0 (4P_z)^0$$

On comparison the symmetry characteristics of composite ring orbitals and metal orbitals. It is fond that the overlap shown in Fig. 5.18. are feasible.

Some workers do not recognize the hybridization process mentioned above and hence, in the diagram, feasibility of overlaps with d^2_z and s metal orbitals is also shown. Also only one ring orbital each of E_1 and E_2 degenerate has been shown in the Figure; the other is at right angles to 'one shown.

The notation cp is used for rings in the Fig. 5.19 thus, cpA_{1g} is actually means composite ring A_{1g} orbitals. From the figure, it is apparent that various types of overlaps can take place. It is widely recognized that $E_{1g} - E_{1g}$ overlap has been very extensive.

As each of these orbitals on one rings as well as metal (here iron) have one electron, such overlap would result in the formation of two covalent bonds by the iron atom and would thus account for the aromaticity of the rings. Other feasible overlaps are as follows (with the direction of shift of electron density indicated by the arrow):–

cpA_{1g} → 4s [or the $(DS)_x$ hybrid orbital]

cpA_{1u} → $4P_z$

cp_{1u} → $4p_x, 4p_y$

cpE_{1g} → $3d_{xy}, 3d_{x^2-y^2}$

($cpA_{1g} - 3d^2_z$) overlap can be avoided if it is presumed that 4s orbital is vacant and all the eight valence electrons are present in the 3d orbitals; in that case $3d^2_z$, by virtue of having an electron pair, becomes ineffective. Similarly, cpA_{1g} – $(DS)_y$ overlap could also be overlooked).

S	–	A_{1g}
P_x P_y	–	E_{1u}
P_z	–	A_{1u}

About the overall bonding in the molecule, two extreme views have been given.

(i) That all interaction, excluding E_{1g}–E_{1g} interactions, have been negligible.

(ii) That E_{1g}–E_{1g} interactions are extensive but some of the other interactions, listed above, may also contribute significantly.

First View

This view includes the Moffitt's hybridization process and also includes the fact that even if the symmetry conditions get satisfied the overlap may not result into any significant bonding, if there is wide disparity in the relative energies of the interacting orbitals. On the basis of this consideration, any interaction involving cpA_{1g} and cpA_{1u} orbitals is rejected as these orbitals, themselves have been of very low energy in comparison to any metal orbital.

Similarly, the E_{2g} and E_{2u} composite ring orbitals, as they are derived from highest energy individual ring E_2 orbitals, have been of very high energy and are not expected to be involved in any meaningful interaction. This leaves only cpE_g orbitals; which can undergo interaction with the d_{xz} and d_{yz} metal orbitals, resulting in two covalent bonds. No doubt, cpE_{1u} orbitals have the right symmetry to overlap $4p_z$ and $4p_y$ metal orbitals, in the metallocenes of the first transition series, but the above said metal orbitals are such high energy (about 80 k cal above 3d) that these overlapping again are not expected to contribute towards metal-ring bonding. These observation can be summarized in the tabular form as in the Table 1.

Table 1

Composite ring orbitals	*Metal orbitals (first transition series)*	*Probability of interaction*
cpA_{1g}	$(DS)_x$	Negligible : cpA_{1g} is of very low energy.
cpA_{1g}	$(DS)_x$	Negligible : cpA_{1g} is of very low energy.
cpA_{1u}	$4p_z$	Negligible : metal orbital is of very high energies.
cpE_{1g}	$3d_{xz}$ $3d_{yz}$	Very high : similar energies
cpE_{1u}	$4p_x$ $4p_y$	Negligible : metal orbital are of very high energy.
cpE_{2g}	$3d_{xy}$, $3d_{x^2-y^2}$	Negligible : cp orbitals are of very high energy.
cpE_{2u}	No metal orbitals of E_{2u} symmetry available.	

The overlapping of double degenerate cpE_{1g} and metal E_{1g} orbitals, gives rise to formation of double degenerate e_{1g} and e^*_{2g} bonding and antibonding metallocene orbitals. In ferrocene, the four electrons (two formerly associated with cpE_{1g} orbitals and two with metal E_{1g} orbitals) get accommodated in the bonding e_{1g} orbitals and e^*_{1g} remain vacant. This accounts for two covalent bonds formed by the metal. As there are two bond covalent, it is evident that one covalent bond is formed with each ring. Such a theory of bonding, based on the involvement of delocalized π-orbitals of the rings, clearly explains why the metal atom behaves as if it gets bonded to the rings as whole and not to any particular carbon atom on the two rings. The formation of one covalent bond with each ring would account for the aromaticity, because the bonding electrons would bring about the number of π-electrons in each of the two rings to six (4n + 2).

The bonding in the various metallocenes is almost the same as described above. The difference in their magnetic properties occurs due to the electron population of these metal orbitals which do not take part in any interaction. The configuration of ferrocene has been depicted in and described in words, along with the configuration of other 3d series metallocenes, below. Based on the works of Moffitt, and Liehr and Ballahusen. It is of interest to not that among 3d elements, manganese does not from compound of the type $(\pi - C_5H_5)_2M$.

$(\pi–C_5H_5)_2Fe$

From Fig. 5.20 it is evident that ferrocene molecule should be diamagnetic, as it has been found to be. Non-bonding metal electrons (six in number) get accommodated in paired state in $(DS)_y$ and E_{2g} orbitals.

$(\pi–C_5H_5)_2Ti$

Titanium atom is having the configuration-$3d^2 4s^2$. Two of these enter $(DS)_y$ and two become E_{1g} electrons. E_{1g} electrons have been involved in the bonding and the $(DS)_y$ electrons, which are non-bonding, remain in the paired state. The molecule is hence, diamagnetic.

$(\pi–C_5H_5)_2V$

Vanadium is having the ground state configuration-$3d^3 4s^2$. In this case, the energy difference between the hybrid $(DS)_u$ orbitals and E_{2g} orbitals is not as large as in case of titanium; rather, it is very small and $(DS)_y$ is almost degenerate with E_{2g} orbitals. Hence, two electrons go to E_{2g} and the remaining three to $(DS)_y$ and double degenerate E_{2g} orbitals in the unpaired state. The metallocene molecule is hence paramagnetic due to the presence of three unpaired electrons.

$(\pi\text{–}C_5H_5)_2Cr$

Number of valence electrons in the chromium atom has been six, one more than the number in vanadium. Two of these enter and one each to two degenerate E_{2g} orbitals. The metallocene molecule is consequently paramagnetic due to the presence of two unpaired electrons.

$(\pi\text{–}C_5H_5)_2Co$

Here the number of valence electrons (in cobalt atoms) has been nine (one more than in iron) Two enter E_{1g}, two enter $(DS)_y$, four enter two E_{2g} orbitals and the ninth one enter $(DS)_x$ orbital in the unpaired state, making the metallocene molecule paramagnetic (one unpaired electrons).

$(\pi\text{–}C_5H_5)_2Ni$

It can be expected that the extra (*i.e.,* tenth) electron of nickel will also enter $(DS)_x$ and would thus pair up with the lone electron present in case of cobalt. However, her $(DS)_x$ is almost degenerate with the antibonding e*1 orbitals and hence, the 10th electron, instead of pairing with the 9th electron present in $(DS)_x$ enters one of the e^*_{1g} orbital. The nickellocene is, consequently, paramagnetic to the tune of presence of two unpaired electrons.

Copper is not able to form such compounds, obviously for the reason, that in the atom only one E_{1g} orbital would be having an unpaired electron, manganese compound is ionic, and so, different from the metallocenes discussed above; this, in a way, reflects the special stability of half filled d sub shell, which the Mn^{2+} ion has, Scandium also forms ionic compound.

Second View

The work of Dunitz and Orgel is the best representative of this view. Dunitz and Orgel also rejected any significant overlapping by cpA_{1u}, cpA_{1u} orbitals on energy grounds, but regards the overlapping by cpA_{1g} and cpE_{2g} orbitals (apart from cpE_{1g} orbitals) as important contribution towards the metal-ring bonding. The main features of their work are as follows :

1. It does not consider any hybridization process.
2. cpA_{1g}, orbital undergoes overlapping with the two metal A_{1g} orbitals to form a_{1g} and a'_{1g} and a''_{1g}-molecular orbitals.

3. cpA_{1g}-metal E_{1g} interaction occurs to result into c_{1g} and c'_{1g} orbitals.
4. cpE_{2g} metal E_{2g} interaction yields c_{2g} and c'_{2g} orbitals. This interaction is responsible for back bonding in the molecule.
5. Position of a'_{1g} and a''_{1g} may vary indifferent metallocenes with respect to e_{2g} and e'_{1g} orbitals.

a'_{1g} is may be nearly degenerate with e_{2g} or sometimes may even be of lower energy. Similarly, a''_{1g} may be of lower of higher energy or degenerate with e'_{2g} orbital. It is the relative energy of these orbitals which decides the magnetic properties of the metallocene molecule.

6. A_{1u}, E_{1u} and E_{2u} composite ring-orbitals can be treated as non-bonding molecular orbitals of the metallocene molecule.
7. The energy level diagram is shown in Fig. 5.21.
8. On the basis of the energy level diagram (Fig. 5.21)., the configurations of various metallocenes can be put as follows :

$(\pi\text{–}C_5H_5)_2$ Ti

$(a_{1g})^2 (A_{1u})^2 (e_{1g})^4 (E_{1u})^4 (a'_{1g})^2$ – Diamagnetic

$(\pi\text{–}C_5H_5)_2$ V

$(a_{1u})^2 (A_{1u})^3 (e_{1u})^4 (a_{1u})^4 (a'_{1g})^2(e_{2g})^2$ – Diamagnetic

Here, a'_{1g} and e_{2g} orbitals are almost degenerate.

$(\pi\text{–}C_5H_5)_2$ Cr

$(a_{1u})^2 (A_{1u})^2 (e_{1g})^4 (E_{2u})^4 (a'_{1g})^2 (e_{2g})^2$ – Paramagnetic (two unpaired electrons)

$(\pi\text{–}C_5H_5)_2$ Fe

$(a_{1g})^2 (A_{1u})^2 (e_{1g})^4 (E_{1u})^4 (a'_{1g})^2 (e_{2g})^4$ – Diamagnetic

$(\pi\text{–}C_5H_5)_2$ Co

$(a_{1g})^2 (A_{1u})^2 (e_{1g})^4 (E_{1u})^4 (a'_{1g})^4 (e_{2g})^2 (a''_{1g})^1$ – Paramagnetic (one unpaired electron)

$(\pi\text{–}C_5H_5)_2$ Ni

$(a_{1g})^1 (A_{1u})^2 (e_{1g})^4 (E_{1u})^4 (a'_{1g})^2 (e_{2g})^4 (e'_{1g})^2$ – Paramagnetic (two unpaired electrons)

Here a''_{1g} has been of higher energy than e_{1g}.

Mono-π-Cyclopentadienyl Metal Complexes

The properties of such compounds, examples of which have been given earlier, indicate that they are also sandwich compounds in which one C_5H_5 ring has been displaced by a group whose π orbital takes part in covalent bonding with the metal. We can take the case of the nickel compound, p-(C5H5)Ni(NO).

NO group is having the configuration $(\sigma)^2$, $(\pi)^4$, $(\pi^*_z)^0$, $(\pi^*_y)^0$. One E_{1g} orbital of the metal is involved in the covalent interaction with the ring, as described earlier, while the other E_{1g} orbitals, again containing a single electron overlaps the π^*_z, orbital of No to establish the second covalent bond.

Cyclopentadienides

These are the ionic compounds which are formed by highly elector 0-positive metals such as sodium-NaC_5H_5), magnesium-Mg $(C_5H_5)_2$ thallium-TIC_5H_5 or scandium and yttirum-$(C_2H_5)_3M$. The metal atom gives away its valence electrons to the ring π-orbitals, forming C_5H_5 ions; the electron obtained by C_5H_5 ring is symmetrically distributed over the whole ring in a delocalized π-orbitals. However, some covalent characteristics are also present in such compounds which arise due to the interaction of metal p orbitals with the E_1 orbitals of the rings.

Manganese also forms anionic compound $Mn(C_5H_5)_2$.

IONIC CYCLOPENTADIENYL COMPOUNDS

Cyclopentadiene is a weak acid ((pKa ~ 20), with is dissociating to give H^+ and $C_5H_5^-$ ion. C_5H_5 forms ionic compounds-cyclopenta-dientides with strongly electropositive metal ions. The more important cyclopentadienides are formed by the alkali metals, MC_5H_5, the alkaline earth metals $M(C_5H_5)_2$, the lanthanides and actinides $M(C_5H_5)_2$, $Mn(C_5H_5)_2$ closely resembles the corresponding magnesium compound and evidently has an ionic structure containing $C_5H_5^-$ and Mn^{2+} ions. The magnetic moment (1.9 B.M) or $Mn(C_5H_5)_2$ is in conformity with the ionic bonding for the above compounds.

The ionic cyclopentadienides have been reported to be very reactive towards air and water and react with ferrous chloride in tetrahedrofuran to yield ferrocene. Some of the ionic compounds of the type $M(C_5H_5)_2$ have like ferrocene, sandwich structures. Hence, the structure of a cyclopentadienyl complex appears to be independent of the bond type.

(c) σ-Cyclo Pental Dienyl Metal Complexes

The transition metal complexes such as $C_5H_5CuP(C_2H_5)_2$ and $(C_5H_5)_3Mn_2(NO)_3$ in which the C_5H_5 group is σ bonded to the metal were first of all reported by Wilkinson and Piper. The copper complex was first of all obtained by the reaction of cuprous oxide with mixture of cyclo pentadiene and triethyl phosphine. Hallam and pauson reported the compound $C_{10}H_{10}Fe(CO)_2$, Wilkinson and Piper independently prepared iron compounds and isoelectronic chromium compound, $C_{10}H_{10}Cr(NO)_2$.

Structure

Infra-red studies of iron and chromium compound reveal multiple frequencies in the C-H stretching region. The N.M.R studies of these compounds by Wilkinson and Piper showed the presence of two distinct groups of proton resonances. One of these occurs in the same region as resonances from other $\pi - C_5H_5$ compounds, the second resonance occurs on the low field side relative to $\pi - C_5H_5$ Peak and is attributed to the protons of the $\sigma - C_5H_5$ ring. The structure proposed for $C_{10}H_{10}Fe(CO)_2$, and $(C_5H_5)_2Cr(NO)_4$ by piper.

The structure was confirmed by Hallam and pauson who obtained confirmatory chemical evidence by isolating two isomers with a benzyl residue in either the $\sigma - C_5H_5$ or the $\pi - C_5H_5$ ring and by showing the rapid hydrolysis of the compound with hydrochloric acid. The structure of $(C_5H_5)_3Mn(NO)_3$ is given as follows. The structure is having a $\sigma -$ and a $\pi - C_5H_5$ ring bond to one of the manganese atoms so as to account for infra-red spectra and diamagnetism of the compound.

2

Interconversion of the Some Hydrides

INTRODUCTION

The methods for preparing boron hydrides give diborane as the main product, it becomes important for its conversion to the other hydrides. Examples of the conditions used to obtain useful yields of higher hydrides are:

(i) Pentaborane is formed when a mixture of diborane and hydrogen is passed through a glass tube at 200-250°C. The yield of this method is good.

$$5B_2H_6 \rightarrow 2B_5H_9 + 6H_2$$

(ii) When pentaborane and hydrogen are heated to 100°C, a mixture of tetraborane and diborane is obtained.

$$2B_5H_6 + 2H_2 \rightarrow 2B_4H_{10} + B_2H_6$$

(iii) A mixture of B_5H_9 and B_5H_{11} is obtained when diborane at a pressure of 120 nun is passed through a tube at 250°C.

$$5B_2H_6 \rightarrow B_5H_9 + B_5H_{11} + 5H_2$$

(iv) By the conversion of one borane into the other : Higher boranes can also be obtained by the conversion of one borane into the other as shown by the following examples

$$B_4H_{10} \xrightarrow{95°C} B_{10}H_{14} \xrightarrow{200°C} B_5H_9 \xrightarrow[\text{disch arge}]{\text{silent electric}} B_{10}H_{16}$$

$$B_4H_{10} \xrightarrow{60°C} B_5H_{11} \xrightarrow{25°C} B_{10}H_{14};\ B_4H_{10} \xrightarrow{200°C} B_5H_9$$

(v) Front B_2H_6. Higher boranes like B_5H_{11}, B_4H_{10}, $B_{10}H_{14}$, B_5H_9, B_5H_{11} can be obtained by heating B_2H_6 at different temperatures as shown below

$$5B_2H_6 \xrightarrow{115°C(U-tube)} 2B_5H_{11} + 4H_2$$

$$2B_2H_6 \xrightleftharpoons{-78 \text{ to } 100°C} B_4H_{10} + H_2$$

$$5B_2H_6 \xrightarrow{100°C\,(sealed\ bulb)} B_{10}H_{14} + 8H_2$$

$$5B_2H_6 \xrightarrow{200-250°C} 2B_5H_9 + 6H_2$$

$$5B_2H_6 \xrightarrow{250°C,\ 120\,mm\,(tube)} B_5H_9 + B_5H_{11} + 5H_2$$

Non-Volatile Boron Hydrides

A large number of non-volatile solid boron hydrides have been prepared by the action of heat on the volatile boron hydrides. For example, diborane on heating to 100°C in a glass vessel yields a colourless film of a non-volatile hydride corresponding to the formula $(BH_{15})_x$, which on further heating changes into a yellow insoluble substance. This is insoluble in carbon disulphide.

When B_4H_{10} is heated, it gives a mixture of yellow hydride (soluble in carbon disulphide) and colourless solid hydride (soluble in carbon disulphide) having a molecular weight of about 140. When B_4H_{10} is stored at room temperature, it undergoes spontaneous decomposition, forming a yellow crystalline hydride having the probable formula $B_{26}H_{36}$.

These non-volatile hydrides need further study.

General Properties of Boron Hydrides

(i) The boranes are volatile and decompose to boron and hydrogen at red heat,

(ii) All the boranes are oxidised by oxygen and explode in air.

(iii) All the boranes or boron hydrides are decomposed by aqueous alkali. Each B-B or B-H bond in a molecule of borane liberates one molecule of hydrogen.

$$B_2H_6 + 6NaOH \rightarrow 2Na_3BO_3 + 6H_2 \uparrow$$

(iv) **Hydrolysis :** All the boranes are decomposed by water liberating hydrogen. The rate of the reaction varies widely. For example,

(a) B_2H_6 is decomposed rapidly.

$$B_2H_6 + 6H_2O \rightarrow 2H_2BO_3 + 6H_2$$

(b) B_4H_{10} and B_5H_{11} react rapidly with water while B_5H_9, B_6H_{10} and $B_{10}H_{11}$ are decomposed only very slowly.

Other examples of hydrolysis of boron hydrides are as follows

$$B_4H_{10} + 12H_2O \rightarrow 4H_3BO_3 + 11H_2$$

$$B_4H_{10} + 4KOH + 4H_2O \rightarrow 4KBO_2 + 11H_2$$

$$B_6H_{12} + 18H_2O \rightarrow 6H_3BO_3 + 15H_2$$

$$B_6H_{12} + 6KOH + 6H_2O \rightarrow 6KBO_2 + 15H_2$$

(v) **Action of NH_3 :** All the boranes react with ammonia, but the products depend upon the conditions.

$$B_2H_6 + NH_3 \xrightarrow[\text{Low temp.}]{\text{Excess } NH_3} B_2H_6.2NH_3$$

$$\xrightarrow[\text{High temp.}]{\text{Excess } NH_3} (BN)_x$$

$$\xrightarrow[\text{High temp.}]{2NH_3.B_2H_6} B_3N_3H_6 \text{ (Borazole)}$$

(vi) **Action of Halogens :** All the boranes react with halogens under controlled conditions.

$$B_2H_6 + Cl_2 \rightarrow B_2H_5Cl + HCl$$

$$B_2H_6 + Cl_2 \rightarrow 3BCl_3 + 3H_2$$

(vii) **Action of Halogen Acids :** All the boranes react with halogen acids. For example, diborane: reacts with HCl in the presence of $AlCl_3$ forming monochloroborane.

$$B_2H_6 + HCl \rightarrow B_2H_5Cl + H_2$$

(viii) **Action of Alkali Metals :** Boranes react with alkali metals. For example, diborane reacts with potassium amalgam forming a mixture of $K_2B_2H_6$, $K_2B_4H_{10}$ etc.

(ix) Alkyl derivatives may be prepared by the reactions of boranes with methyl iodide or in general alkyl halide.

$$B_2H_6 \xrightarrow{4CH_3I} B_2H_2(CH_3)_4 + 4HI$$

(x) Lewis bases as CH_3CN, C_5H_5N and $(CH_3)_2S$ react with decaborane to liberate hydrogen.

$$2CH_3CN + B_{10}H_{14} \rightarrow B_{10}H_{12},2CH_3CN + H_2$$

(xi) **Formation of Addition Compounds :** Since all the boranes are electron-deficient compounds, these: form addition compounds (or adducts) with a variety of donor molecules.

Uses : Diborane is used

1. As a catalyst for polymerisation reactions.
2. As a reducing agent for organic compounds.
3. For welding torches.
4. For conversion into substances which are used as high energy fuels and propellants. In the near future, other hydrides of boron may find certain uses. Recently, boron hydrides have acquired more importance by using them as potential high energy rocket fuels

$$B_2H_6 + 3O_2 \rightarrow B_2O_3 + 3H_2O; \quad \Delta H = -2025 \text{ kJ/mole.}$$

(B) Preparation of Individual Boron Hydrides

Let us discuss the various methods of preparation of different hydrides of boron.

DIBORANE

Preparation

(i) The reaction of boron trichloride or bromide with hydrogen over Cu - Al catalyst yields the diborane :

$$2BCl_3 + 6H_2 \rightarrow B_2H_6 + 6HCI$$

$$2BBr_3 + 6H_2 \rightarrow B_2H_6 + 6HBr$$

The yield of this method is poor.

(ii) It may by prepared by dropping an ethereal solution of boron trifluoride on sodium borohydride.

$$3NaBH_4 + 4BF_3 \rightarrow 3NaBF_4 + 2B_2H_6$$

(iii) It can also be prepared by the action of dilute H_2SO_4 on sodium borohydride.

$$2NaBH_4 + H_2SO_4 \rightarrow B_2H_6 + 2H_2 + Na_2SO_4$$

(iv) It is also prepared by the reaction of boron trifluoride ether complex with a refluxing suspension of lithium or sodium hydride.

$$6LiH + 8(C_2H_5)_2O.BF_3 \rightarrow 6LiBF_4 + B_2H_6 + 8(C_2H_5)_2O$$

(v) It is prepared by the reaction of boron trichloride with lithium aluminium hydride in ether solution.

$$3LiAlH_4 + 4BCl_3 \xrightarrow{\text{Ether}} LiCl + 3AlCl_3 + 2B_2H_6$$

This method gives a quantitative yield.

Properties of Diborane

1. It is an inflammable colourless gas having a sticky sweet odour and is extremely toxic. It is an extremely reactive gas and hence must be handled in a special apparatus so that it comes in contact only with glass and mercury.
2. **Decomposition**

(a) **By heat :** It is stable at low temperature only in the absence of moisture. At higher temperatures B_2H_6 is decomposed into a number of higher boranes. For example

$$5B_2H_6 \xrightarrow{115°C(U-tube)} 2B_5H_{11} + 4H_2$$

$$2B_2H_6 \xrightarrow{-78 to 100°} B_4H_{10} + H_2$$

$$5B_2H_6 \xrightarrow{100°C(sealed bulb)} B_{10}H_{14} + 8H_2$$

$$5B_2H_6 \xrightarrow[(tube)]{250°C, 120 mm} B_5H_9 + B_5H_{11} + 5H_2$$

$$5B_2H_6 \xrightarrow{200-250°C} 2B_5H_9 + 6H_2$$

At red heat B_2H_6 is decomposed into boron and H_2.

$$B_2H_6 \xrightarrow{\text{Red heat}} 2B + 3H_2$$

(b) **By water :** Hydrolysis. It is readily decomposed by water into boric acid and H_2.

$$B_2H_6 + 6H_2O \rightarrow 2H_3BO_3 + 6H_2$$

(c) **By aqueous alkalies :** *Hydrolysis.* When B_2H_6 is allowed to pass into a concentrated aqueous solution of KOH at 0°C, reaction takes place and the solution obtained on evaporation in vacuum deposits potassium hypoborate, $K_2(BH_2OH.BH_2OH)$ or $K_2(B_2H_6O_2)$.

$$B_2H_6 + 2KOH \xrightarrow{0°C} K_2(B_2H_6O_2) \quad ...(i)$$

If KOH solution is diluted, then potassium hypoborate, $K_2(B_2H_6O_2)$ gets converted into potassium metaborate, KBO_2.

$$K_2(B_2H_6O_2) + 2H_2O \rightarrow 2KBO_2 + 5H_2 \quad ...(ii)$$

Thus the hydrolysis of B2H6 by a dilute solution of KOH can be shown by the following equation which is obtained by combining equations (i) and (ii)

$$B_2H_6 + 2KOH + 2H_2O \rightarrow 2KBO_2 + 5H_2$$

The evolution of H_2 by the action of aqueous alkalies makes B_2H_6 as a useful reducing agent.

3. **Action of lithium ethyl (LiC_2H_5) and alkali metal hydrides :** LiC_2H_5 and alkali metal hydrides suspended in ether react with B_2H_6 and produce alkali metal borohydrides, MBH_4.

$$2B_2H_6 + 2LiC_2H_5 \rightarrow 3LiBH_4 + B(C_2H_5)_3$$
$$B_2H_6 + 2LiH \rightarrow 2LiBH_4$$
$$B_2H_6 + 2NaH \rightarrow 2NaBH_4$$

4. **Reaction with trimethyl aluminium, $Al(CH_3)_3$:** Aluminium borohydride, $Al(BH_4)_3$ is formed.

$$Al(CH_3)_3 + 2B_2H_6 \rightarrow B(CH_3)_3 + Al(BH_4)_3$$

5. **Action of O_2 :** *Ignition or combustion.* Pure B_2H_6 undergoes no change when mixed with dry air or O_2 at room temperature, but it may ignite (*i.e.,* burn), if impure. In this reaction a large amount of energy gets liberated.

$$B_2H_6 + 3O_2 \rightarrow B_2O_3 + 3H_2O + 2025 \text{ kJ}$$

The production of large amount of heat in the above reaction makes B_2H_6 a useful rocket fuel.

6. **Action of halogens :** It reacts with halogens (not iodine) under different conditions to yield different products.

$$B_2H_6 + 6Cl_2 \xrightarrow{25°C} 2BC1_3 + 6HCl$$
(vigorous reaction)

$$B_2H_6 + Br_2 \xrightarrow{100°C} B_2H_5Br + HBr$$ *(slow reaction)*

7. **Action of halogen acids :** B_2H_6 reacts readily with HCl acid in presence of $AlCl_3$ as catayst to form chlorodiborane (B_2H_5Cl).

With HBr it reacts at 90°C in presence of $AlBr_3$ to give B_2H_5Br while with HI at 50°C in the absence of a catalyst to give B_2H_5I.

$B_2H_6 + HX(X = Cl, Br, I) \rightarrow B_2H_5X + H_2$

8. **Action of sodium or potassium amalgam :** When B_2H_6 is treated with sodium or potassium amalgam (reducing agents) in cold, crystalline product of the composition, B_2H_2 2Na or B_2H_6.2K is formed.

2Na *(in the form of amalgam)* + $B_2H_6 \rightarrow B_2H_6.2Na$

2K *(in the form of amalgam)* + $B_2H_6 \rightarrow B_2H_6.2K$

9. **Reaction with NH_3 :** Diborane reacts with NH_3 to form various products depending on the experimental conditions. For example.

(a) When B_2H_6 reacts with excess of NH_3 at low temperature (– 120°C), it forms an addition compound called di-ammoniate of diborane, $B_2H_6.2NH_3$ which yields a conducting solution in liquid ammonia.

$B_2H_6 + 2NH_3 \text{ (Excess)} \xrightarrow{\text{Low temp } (-120°C)} B_2H_6.2NH_3$

$B_2H_6.2NH_3$ is a white non-volatile solid. It is soluble in water. It is an ionic compound and hence is represented as $[BH_2(NH_3)_2]^+ [BH_4]^-$.

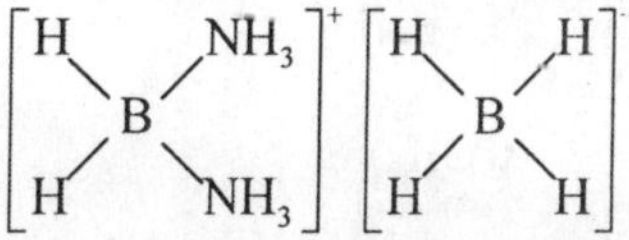

The formation of $B_2H_6.2NH_3$ is the only case of non-symmetrical cleavage of B_2H_6 by consecutive substitution.

(b) With excess of NH_3 and at high temperature, boron nitride, $(BN)_X$ gets formed.

$xB_2H_6 + 2xNH_3 \text{ (Excess)} \xrightarrow{\text{High temp.}} 2(BN)_x + 6xH_2$

(c) When B_2H_6 and NH_3 are treated at high temperature in 1 : 2 ratio, borazole, $B_3N_3H_6$ (also called borazine) is formed.

$3B_2H_6 + 6NH_3 \xrightarrow{\text{High temp.}} 2B_3N_3H_6 + 12H_2$

Borazole is having a flat hexagonal ring structure containing-BH= and -NH=groups alternately.

Thus, it has an analogous electronic structure to that of C_6H_6 and is, therefore coiled inorganic benzene.

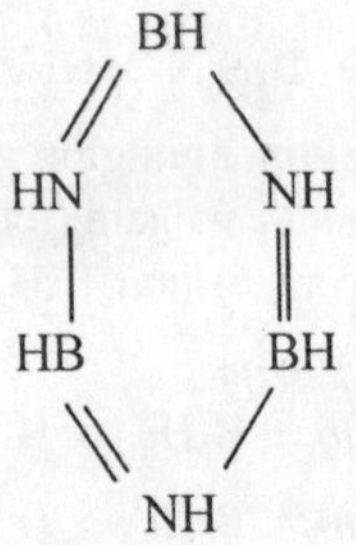

Borazole ($B_3N_3H_6$)

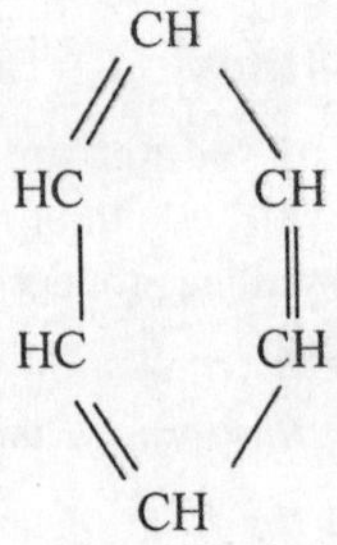

Benzene (C_6H_6)

10. **Formation of adducts :** Due to its electron-deficient nature, B_2H_6 reacts with a number of molecules having lone pairs of electrons [*e.g.*, $(CH_3)_2O$, $(CH_3)_2S$, CO, PF_3, $N(CH_3)_3$ etc.] (Lewis bases) to form adducts or addition compounds. Hence in these reactions B_2H_6 acts as an acceptor (Lewis acid) and the molecules such as (CH3)20, (CH3)2S etc. act as donors (Lewis bases).

Acceptor (Lewis acid)	*Donor (Lewis bases)*		*Adduct or addition compound*
$\frac{1}{2}B_2H_6$ +	$(CH_3)_2O$ Ether	$\rightarrow$	$(CH_3)_2O \rightarrow BH_3$ Ether borane
$\frac{1}{2}B_2H_6$ +	$(CH_3)_2S$ Thioether	$\rightarrow$	$(CH_3)_2S \rightarrow BH_3$ Thioether borane
$\frac{1}{2}B_2H_6$ +	CO	$\xrightarrow[28\,atm]{200°C}$	$CO \rightarrow BH_3$ Carbonyl borane

11. **Methylation of B_2H_6 with trimethyl borane, $B(CH_3)_3$:** B_2H_6 reacts at ordinary temperature with $B(CH_3)_3$ to form a series of methyl derivatives of B_2H_6 which are called methyl boranes. No more than four H atoms in B_2H_6 molecule could be replaced by CH_3 groups or any alkyl groups as depicted below.

    The above reactions clearly reveal that none of the bridge H-atoms in B_2H_6 has been replaced by CH_3 group *i.e.*, in the reaction between B_2H_6 and $B(CH_3)_3$, both three-centre B-H-B bonds remain undisturbed.

12. **Reactions with PH_3 and its derivatives**

(a) At ordinary temperature B_2H_6 reacts with PH_3 to yield a non-volatile white solid having the composition $(PH_2BH_2)A$.

$$\frac{x}{2}B_2H_6 + xPH_3 \xrightarrow{-xH_2} (PH_7BH_2)_x + xH_2$$

(b) At – 110°C B_2H_6 reacts with PH_3 to form an unstable product with the composition $PH_3.BH_3$.

$$\frac{1}{2}B_2H_6 + PH_3 \xrightarrow{-110°C} .PH_3.BH_3$$

(c) When $(CH_3)_2$ PH reacts with B_2H_6, $(CH_3)_2$ Ph.BH_3 is formed which, on being heated at 150°C gives P_3 $(CH_3)_6$ $(BH)_3$ which is a cyclic compound, because it has ring structure like C_6H_6.

$$\frac{1}{2}B_2H_6 + (CH_3)_2PH_3 \rightarrow (CH_3)_2\ PH.BH_3$$

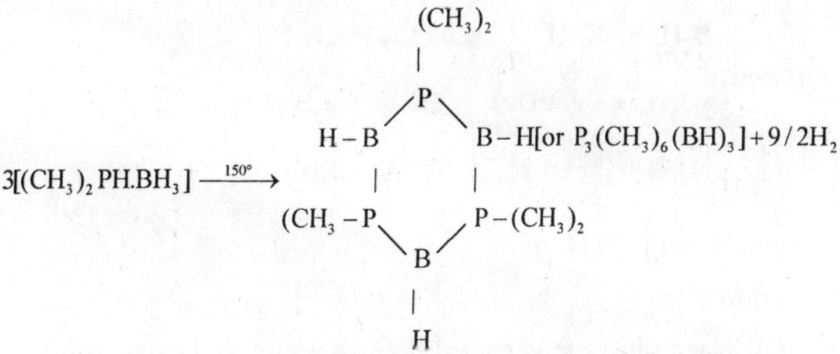

13. **Reaction with alkali metals (M) in liquid NH_3 :** The reaction is believed to take place as follows :

$$B_2H_6 + M + NH_3 \rightarrow M[BH_4] + BH_2.NH_2 + 1/2\ H_2$$

Tetraborane-10 (B_4H_{10})

Preparation

(i) It is obtained by Stock's method by acidolysis of magnesium boride.

(ii) When diborane is passed through a hot-cold reactor by maintaining at 120°C and then at 10°C, tetraborane is formed.

(iii) B_5H_{11} and hydrogen at 100°C give tetraborane and diborane as the main products.

Properties

It is a colourless gas and poisonous gas with a disgusting odour. This can be readily condensed to a colourless liquid with melting point = –120°C and boiling point = 18°C. It is non-inflammable in dry air at room temperature.

Pentaborane-9 (B_5H_9)

When a mixture of diborane and hydrogen is circulated through a glass tube at 200-250°C, pentaborane is obtained in a good yield.

$$B_2H_6 + H_2 \xrightarrow{200-250°C} B_5H_9$$

Properties

Some of the reactions shown by B_5H_9 are as follows

$$B_5H_9 + C_2H_4 \xrightarrow{AlCl_3} B_5H_8C_2H_5$$

$$B_5H_9 + 2LiH \xrightarrow{\text{Diethyl ether}} Li_2B_5H_{11}$$

$$B_5H_9 + 2Na(Hg) \xrightarrow{\text{ether}} Na_2B_5H_9 \xrightarrow{\text{Vacuum}} B_5H_9$$

$$B_5H_9 + 2Li \xrightarrow{\text{Liq. } NH_3} Li_2B_5H_9$$

Pentaborane-11 (B_5H_{11})

Preparation

(i) When diborane is passed through a tube at 115°C, a moderate yield of the unstable pentaborane, B_5H_{11}, is obtained.

(ii) Pentaborane, B_5H_{11} is also obtained by heating a mixture of diborane and tetraborane under suitable conditions.

$$B_2H_6 + 2B_4H_{10} \rightarrow 2B_5H_{11} + 2H_2$$

Properties

(i) In presence of weak Lewis bases like ethers, amines etc. B_5H_{11} yields B_6H_{10} and B_2H_6

$$2B_5H_{11} \xrightarrow[\text{a Lewis base}]{\text{In presence of}} B_6H_{10} + 2B_2H_6$$

(ii) B_5H_{11} reacts with CO in a hot-cold flow reactor to form the carbonyl adduct, B_4H_8CO.

$$B_5H_{11} + 2CO \rightarrow B_4H_5CO + BH_3CO$$

(iii) When heated with H_2 at 100°C it yields B_4H_{10}.

Hexaborane B_6H_{10}

(i) It is obtained in 60% yield when magnesium boride is treated with 8 M orthophosphoric acid

(ii) When B_5H_{11} is decomposed in the presence of weak lewis bases such as diethyl ether, hexaborane is produced along with diborane.

$$2B_5H_{11} \rightarrow B_6H_{10} + 2B_2H_6$$

Enneaborane B_9H_{15}

(i) It is prepared by the reaction between liquid pentaborane and gaseous diborane at 25 atm pressure.

(ii) It is obtained on a large scale by the reaction of B_5H_{11} on the surface of crystalline hexamethylene tetramine.

Decaborane-14 (B_{10} H_{14})

When diborane is heated at about 100°C, decaborane is obtained.

Some reactions shown by this hydride are :

$$B_{10}H_{14} + NaH \xrightarrow{(C_2H_5)_2O} NaB_{10}H_{13}.(C_2H_5)_2O + H_2$$

$$B_{10}H_{14} \xrightarrow[\text{liq. } NH_3]{2Na^+ \text{ in}} (Na^+)_2[B_{10}H_{14}]^{2-}.$$

$$B_{10}H_{14} + 2L\ (L{=}CH_3CN,\ R_3N,\ R_3P) \xrightarrow{\Delta} B_{10}H_{12}L_2 + H_2$$

$$B_{10}H_{14} + CH_3MgI \rightarrow B_{10}H_{13}MgI + CH_4$$

A large number of other hydrides are also known which have been studied. These are

(i) Triborane, B_3H_8

(ii) Octaborane, B_8H_{12}

(iii) Decaborane, $B_{10}H_{16}$

(iv) Octadecaborane, $B_{18}H_{22}$.

It is not possible to discuss these hydrides in this short chapter.

Structure of Hydrides of Boron

The structures of boron hydrides have been determined mainly by experiments on X-rays diffraction, electron diffraction and nuclear magnetic resonance, and the bonding of the atoms has been interpreted by the wave-mechanics. The structures assigned to the boron hydrides are not derived from their chemical properties.

Structure of Diborane

The structural problem, which has been a source of controversy for many years, can be tackled into two stages, first we must consider the relative positions of the atomic nuclei concerned *i.e.,* the nuclear configuration, and secondly we must give a theoretical interpretation of the bonding in a particular nuclear configuration.

1. Nuclear Configuration of Diborane

Of the many structures proposed, consideration may be given to :

(a) ionic structures,

(b) ethane-like structures, and

(c) hydrogen-bridge structures.

Let us discuss these one by one:

(a) **Ionic Structures :** These structures were based upon the faulty assumption that reactions of diborane with ammonia depended upon the presence of acidic hydrogens in diborane. Diborane was assumed to have formulation both as a dibasic acid.

$$H_2^{2+}\left[\begin{array}{c} H\ \ H \\ H:\ddot{B}:\,:\ddot{B}:H \end{array}\right]^{2-}$$

and as monobasic and

$$H^{+}\left[\begin{array}{c} H\ \ H \\ H:\ddot{B}:\,:\ddot{B}:H \\ \ddot{H} \end{array}\right]^{-}$$

Ionic structure could not be supported by experimental data.

(b) **Ethane-like Structures :** The molecular formula of diborane is B_2H_6. It does not undergo thermal dissociation up to 100°C, It is diamagnetic and hence the diborane contains no unpaired electrons.

The molecule of diborane is electron deficient as it has got 12 valency electrons as compared to the corresponding ethane which has 14 electrons, all of which are shared pairs.

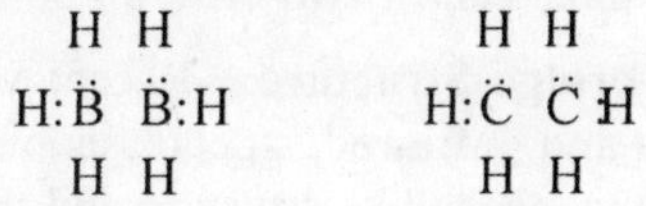

Diborane Ethane

In diborane, no electrons are left between the two boron atoms to link the two.

According to Sidgwick, diborane possesses ethane-like structure with the difference that two of hydrogen atoms in diborane are held by one electron bond (1) as

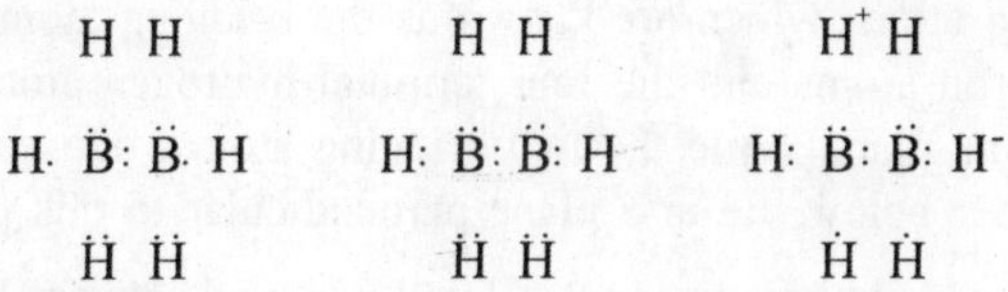

I II III

As the electronegativities of hydrogen and boron are nearly the same, this fact is supporting the possibility of existence of such one-electron bonds. This structure is further stabilised by resonance of the one electron hydrogen-boron bonds among all six positions. It is further suggested that resonance might also involve structures having zero-electron bonds (II) and (III). All the three structures (I, II and III) were wrongly supported by Baeuer's early wrong interpretation of electron diffraction data. As there was not enough support for one-electron bonds, the ethane-like structure of diborane was rejected.

(c) **VB Method :** The earliest attempt at rationalizing the dimerization of borane invoked resonance in a valence bond (VB) context; Although adequate from a formal point of view, it suffers from the usual unwiledness of VB terminology when extensive delocalization exist.

(d) **Ethylene Isoelectronic Structure :** Another attempt considers the $B_2H_4^{2-}$ anion as isoelectronic and isostructural with ethylene, C_2H_4. Such an ion would have a cloud of electron density above and below the B–H plane. The neutral B_2H_6 molecule could then be formally produced by embedding a proton in the electronic cloud above and below the plane of the $B_2H_4^{2-}$ ion. Although, this may appear to be somewhat farfetched, it is but a simplistic way of describing the bonding model which is currently accepted as best-the three-center, two-electron bond.

(e) **Hydrogen-bridge Structure :** Recent work on the electron diffraction and infra-red, spectroscopic data conclusively eliminates the ethane-like structure and the ionic structure but supports the hydrogenbridge structure in which two hydrogen atoms form a bridge between the two boron atoms. The terminal B–H bonds are normal electron pair bonds. The various bond distances and angles.

(i) Four hydrogen atoms, two on the left and two on the right, known as terminal hydrogens, of are in different environment front other two atoms which are known as the bridging atoms. The two boron atoms and the four terminal hydrogen atoms lie in the same plane while the two bridging atoms, one above and the other below, lie in a plane perpendicular to this plane.

(ii) Bond lengths between B...H–is 1.33Å and between B-H is 1.19Å which shows that electron deficiency is associated with the bridge atoms.

The bridge structure discussed above is supported by the following additional facts :

(i) **Electron-diffraction measurements :** These measurements are very useful but the results are the quite difficult to have interpretation. Recently, Hedberg and Schomaker studied the diffraction results of boron hydrides and they supported the bridge structure with the dimensions. A further study by Bartell and Caroll confirmed this structure for both diborane and deuterodiborane, with terminal and normal B–H linkages of 1.12Å and 1.334Å respectively. This study also confirms that boron atoms and the four terminal hydrogen atoms are co-planar and the two bridge hydrogen atoms occupy the place above and below the plane.

(ii) **Vibrational Spectral Studies :** The Raman and infra-red vibrational spectra of diborane support the bridge structure. If the ethane model of diborane is considered to be correct, it will only provide one symmetrical stretching vibration but the bridge structure suggests two, *i.e.,* one due to stretching of the B–H bonds in the bridges and the other due to the stretching of terminal B--H bonds. Experimentally, the two intense frequencies (2102 and 2523 cm^{-1}) have been obtained in the Raman spectrum which could be assigned to two bond-stretching vibrations in the bridge structure of diborane. Thus, the Raman spectra support the bridge structure and reject the ethane-like structure of diborane.

(iii) **Specific heat measurement :** The experimental specific heat of diborane is found to be 54.6 kJ. If the ethane model is considered, it should be only 12.6 kJ. But if the bridge structure is considered, it should be 21 – 63kJ. Thus, a much agreement results if a bridge model is considered.

(iv) **N.M.R. studies :** This provides the most convincing evidence for the diborane structure. When N.M.R. of isotopically labelled diborane "B_2H_6" is obtained, three main regions of absorption are obtained, one associated with the boron atoms, one with the terminal hydrogen atoms and the one with the bridge hydrogen atoms. The fine structure of these absorption bands, which results from interactions between neighbouring nuclei is the same what is expected from the bridge model.

(v) **Methylation Experiment :** Diborane cannot be methylated beyond $(CH_3)_4B_2H_2$. The hydrogens, which cannot be methylated evidently, are in a different environment. These are the bridge hydrogen atoms.

Nature of Bonding in Hydrogen-bridge Structure of Diborane

The physical evidences discussed above show quite definitely that the nuclear configuration of diborane is bridge-like, and we must now try to explain the nature of bonding in such a molecule. The terminal B–H bonds are of same length as expected for normal single B-H bonds and are, accordingly, assumed to be of this type. In these circumstances the electron deficiency must be in the bridge bonds, and we shall now interpret this hydrogen-bridge structure.

Three-Centre or Banana Bond Structure

Longuet gave a description of diborane in which two BH_2 groups are linked by two equivalent 'three-centre' or 'banana' bonds, each of which embraces the two boron atoms and one of the bridge hydrogens. However, the formation of these bonds can be explained as follows:

The atomic number of boron is five and hence its electronic configuration is $1s^2\ 2s^2\ 2p^1$. In diborane molecule, each boron atom in the excited state has initial uncoupling of the paired 2s electrons and promotion of one of the 2s electrons to $2p_y$ orbital. Then, there are three unpaired electrons.

	1s	2s	$2p_x$	$2p_y$	$2p_z$
Boron atom (ground state)	↑ ↓	↑ ↓	↑		
Boron atom (excited stated)	↑ ↓	↑	↑	↑	

Fig. 1

Hybridisation of three half-filled orbitals (2s, $2p_x$ and $2p_y$) together with one remaining empty orbital ($2p_z$) gives four sp^3 hybrid orbitals. In these four hybridised orbitals, three contain unpaired electrons whereas fourth orbital is empty. In diborane formation, is orbital of a bridging hydrogen atom is overlapping with two sp^3 hybrid orbitals, one from each boron atom, to form a three-centre bond containing one pair of electrons but embracing all the three nuclei, *i.e.,* B–H–B. This represents the formation of one of the bridges. Similarly, one can represent the formation of a second three-centre bond.

In diborane, there are six hydrogen atoms. Two hydrogen atoms are used for the formation of two three-centre bonds whereas remaining four hydrogen atoms are joined to two boron atoms, *i.e.,* two are attached to each boron atom. Two sp^3 hybrid orbitals of each boron atom overlap with 1s orbitals of two hydrogen atoms forming two sigma (σ) bonds. The orientation of four hydrogen atoms about each boron atom is approximately tetrahedral. A three-centre bond has a banana shape. This is due to the repulsion between the positive charges on the two bridge hydrogen atoms, causing the three-centre bonds to he bent away from each other in the middle.

Simple Representation of Diborane Molecule

Represents diborane molecule. Dark circular spot (•) represents electrons present in boron atoms, empty spot (O) represents electron present in the hydrogen atom. The two dotted loops represent the three-centred bond. Each loop linking up three atoms (two boron atoms and one hydrogen atom) encloses only two electrons.

MATHEMATICAL PICTURE OF STRUCTURE OF BORON HYDRIDES

Consider each boron atom to he sp^3 hybridized. The two terminal B–H bonds on each boron atom presumably are simple σ bonds involving a pair of electrons each. This accounts for eight of the total of twelve electrons available for bonding. Each of the bridging B–H–B linkages then involves a delocalized or three-center bond as follows. The appropriate combinations of the three orbital wave functions, ϕ_{B1}, ϕ_{B2} (approximately sp^3 hybrids), and ϕ_{11}, (an s orbital) result in three molecular orbitals:

$$\psi_b = \frac{1}{2}\phi_{B2} + \frac{1}{\sqrt{2}}\phi_{11} \qquad ...(1)$$

$$\psi_n = \frac{1}{\sqrt{2}}\phi_{B1} + \frac{1}{\sqrt{2}}\phi_{B1} \qquad ...(2)$$

$$\psi_a = \frac{1}{2}\phi_{B1} + \frac{1}{2}\phi_{B2} + \frac{1}{\sqrt{2}}\phi_{11} \qquad ...(3)$$

where ψ_b, is a bonding MO, ψ_a is an antibonding Mo, and ψ_a is, to a first approximation, a nonbonding MO. The diagrammatic possibilities of overlap together with sketches of the resulting MOs and their relative energies.

Each bridging bond thus consists of a bonding MO containing two electrons. Although the nonbonding orbital could conceivably accept an additional pair of electrons, this would not serve to stabilize the molecule beyond that achieved by the configuration ψ_b^2. The second B–H–B bridge likewise may be considered to have a configuration ψ_b^2. This accounts for the total of twelve bonding electrons and provides the rationale for the existence of the dimer.

Diborane provides examples of two types of bonds found in higher boranes : the two-center, two-electron B–H terminal bond and the three-center, two-electron, bridgine B–H–B bond. Two other bonds are of importance in the higher analogues :

APPLICATIONS OF LOCALISED BONDING AND MOLECULAR ORBITAL THEORY OF POLYHEDRAL BORANES

The structures given in earlier figures show lines joining each pair of boron atoms which are geometrical neighbours. A relationship between this purely geometrical connectivity and localised bonding has been attempted by Lipcomb. Assuming that a borane of formula $B_n H_{n+m}$ may contain BB, BH, BH_2, two it centre bonds and BBB and BHB there centre bonds, the following rules form a basis for assigning a localised bonding structure:

(i) Every pair of boron atoms which are geometrically neighbours must be connected by at least one BB, BHB or BHB or BBB bond.

(ii) Each boron atom, since it has four valence atomic orbitals participate in (4-x-y) boron framework bonds, where x is the number of bridge hydrogen atoms bonded to a particular boron atom.

(iii) No two boron atoms may be bonded together by both a two-centre and a three-centre (BBB or BHB) bond.

(iv) The electron balance for boron atoms is

$$2\ \Sigma BHB + 2\ \Sigma BB + m + n = 3n \qquad \text{... (1)}$$

The hydrogen atom balance is $\Sigma BB + \Sigma BH_2 + n = n + m$ ... (2)

The total number of three centre bonds will be equal to the number of boron atoms, *i.e.*,

$$\Sigma BBB + \Sigma BHB = n \qquad \text{... (3)}$$

On combining equations (1), (2) and (3), we find that the molecule $B_n H_{n+m}$ must have

$$(\Sigma BHB - \frac{1}{2}m) \text{ BB bonds and } (n - \Sigma BCB) \text{ BBB bonds.}$$

On the basis of above rules, the topological structures of B_4H_{10}, B_5H_9 and $B_{10}H_{14}$.

C. Polymeric Hydrides

In polymetric hydrides, two or more metal atoms are linked by hydrogen bridges. Examples are the complex hydrides of the type:

Lithium Aluminium Hydride, $LiAlH_4$: It was discovered by Finholt, Bond and Schlesinger in 1947.

Preparation : It is best prepared by the action of aluminium chloride on lithium hydride in ether containing a small amount of lithium aluminium hydride which acts as a promotor.

$$3LiH + AlH_3 \rightarrow 3LiCl + AlH_3$$

$$LiH + AlCl_3 \rightarrow LiAlH_4$$

$$4LiH + AlCl_3 \xrightarrow{Ether} LiAlH_4 + 3LiCl \downarrow$$

The lithium aluminium hydride remains in solution in ether whereas lithium chloride is insoluble in it. The precipitated lithium chloride is filtered off, the ether is evaporated and lithium aluminium hydride remains in the solution. If the solution is now allowed to stand before filtration and then evaporated, lithium aluminium hydride of up to 99% purity can be obtained.

Properties

(i) It is a colourless solid which is stable to air in the cold.

(ii) When it is heated, it decomposes according to the equation.

$$2LiAlH_4 \rightarrow 2LiH + 2Al + 3H_2\uparrow$$

(iii) It reacts so vigorously with a small quantity of water that it becomes incandescent.

$$LiAlH_4 + 4H_2O \rightarrow LiOH + Al(OH)_3 + 4H_2\uparrow$$

(iv) Its solution in ether has proved a most powerful reducing agent. For, example, it reduces aldehydes to alcohols and gives high yield.

$$4RCHO + LiAlH_4 \rightarrow (R\text{–}CH_2O)_4LiAl$$

$$\uparrow 4H_2O$$

$$RCH_2OH + LiOH + Al(OH)_3$$

In general, it is used as a reducing agent almost for all the substances containing a carbonyl group $>C=O$, such as ketones, carboxylic acids and their derivatives to alcohols. If a double bond is present in the molecule of the aldehyde, it is not affected during reduction by this reagent.

(v) It is used for the preparation of hydrides of silicon, tin, boron, aluminium, germanium, etc.

$$LiAlH_4 + SiCl_4 \xrightarrow{Ether} SiH_4 + LiCl + AlCl_3$$

$$LiAlH_4 + SnCl_4 \xrightarrow{Ether} SnH_4 + LiCl + AlCl_3$$

$$3LiAlH_4 + 4BCl_3 \xrightarrow{Ether} 2B_2H_6 + 3LiCl + 3AlCl_3$$

$$3LiAlH_4 + AlCl_3 \xrightarrow{\text{Ether}} 4AlH_3 + 3LiCl$$

$$LiAlH_4 + 3B_2H_6 \xrightarrow{\text{Ether}} LiBH_4 + Al(BH_4)_3$$

$$LiAlH_4 + Zn(CH_3)_2 \rightarrow ZnH_2 + LiAlH_2(CH_3)_2$$

$$LiAlH_4 + 4RNH_2 \rightarrow [LiAl(RNH)_4] + 4H_2 \uparrow$$

Uses

(i) It is used as a reducing agent for compounds containing keto group,

(ii) It is used for the preparation of hydrides of a number of metals like tin, boron, aluminium, germanium, etc.

Structure : $LiAlH_4$ is assumed to have a tetrahedral $(AlH_4)^-$ ion. Thus, its structure is

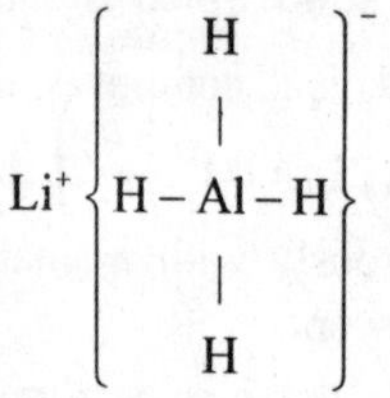

Lithium Borohydride, $LiBH_4$

Preparation

(i) It is prepared by the action of sodium borohydride on lithium chloride.

$$LiCl + NaBH_4 \rightarrow LiBH_4 + NaCl$$

(ii) It may be prepared by the action of diborane on lithium ethyl or hydride suspended in ether.

$$3B_2H_6 + 3LiC_2H_5 \rightarrow 3LiBH_4 + B\ (C_2H_5)_3$$

$$B_2H_6 + 2LiH \rightarrow 2LiBH_4$$

(iii) It may be prepared by the action of aluminium borohydride on lithium ethyl in benzene.

$$Al(BH_4)_3 + 3LiC_2H_5 \rightarrow 3LiBH_4 + Al(C_2H_5)_3$$

Properties

(i) It is a white solid which melts at 275°C with the evolution of hydrogen.

(ii) It reacts with methyl alcohol down to –100°C.

$$LiBH_4 + 4CH_3OH \rightarrow LiOCH_3 + B(OCH_3)_3 + 4H_2$$

and with hydrogen chloride down to –80°C.

$$2LiBH_4 + 2HCl \rightarrow 2LiCl + B_2H_6 + 2H_2.$$

(ii) It reacts with iodine at 125°C to form BI_3.

Structure : The crystal of LiBH4 has an orthorhombic lattice. Each Li^+ ion is associated with four tetrahedral BH_4 ions. $LiBH_4$ is not isomorphous with $LiAlH_4$, $LiClO_4$, $LiBF_4$, $MgSO_4$, etc.

Hydrides of Silicon or Silanes

Silicon forms a number of hydrides which are known as silanes. These are similar to the paraffin hydrocarbons in composition. When magnesium silicide is treated with dilute HCl, a mixture of hydrides is obtained which were separated and studied by Stock.

[SiH_4, Si_2H_6, Si_3H_8, Si_4H_{10}, Si_5H_{12}, Si_6H_{14}, $(Si_2H_4)_n$, $(Si_2H_2)_n$, etc.]

$$Mg_2Si + 4HCl \rightarrow 2MgCl_2 + SiH_4.$$

Silane

Preparation : It is best prepared by passing dry hydrogen gas over hot silicon in an electric arc.

$$Si + 2H_2 \rightarrow SiH_4$$

It may be prepared by the action of $LiAlH_4$ on $SiCl_4$ in ether solution

$$SiCl_4 + LiAlH_4 \rightarrow SiH_4 + AlCl_3 + LiCl$$

Properties : It is a gas having a faint odour. It breaks down into its elements at 400°C.

$$SiH_4 \xrightarrow{400°C} Si + 2H_2$$

When silane is passed into the silver salt, it precipitates silver.

$$SiH_4 + AgNO_3 \rightarrow Si + 4Ag + 5HNO_3$$

With copper sulphate, it gives a ppt. of copper silicide (Cu_2Si),

$$2CuSO_4 + SiH_4 \rightarrow Cu_2Si + 2H_2SO_4$$

Disilane

Preparation : It is prepared by the action of conc. HCl acid on lithium silicide (Li_6Si_2)

$$Li_6Si_2 + 6HCl \rightarrow 6LiCl + Si_2H_6$$

Properties : It is a colourless liquid which decomposes into its elements at 200°C.

$$Si_2H_6 \rightarrow 2Si + 3H_2$$

With alkalies, it reacts to yield silicates with the liberation of hydrogen.

$$Si_2H_6 + 4KOH + 2H_2O \rightarrow 3K_2SiO_3 + 7H_2$$

With chloroform and carbon tetrachloride, it reacts to form silicon tetrachloride.

$$Si_2H_6 + 3CHCl_3 \rightarrow 2SiCl_4 + HCl + 4H_2 + 3C$$

$$Si_2H_6 + 2CCl_4 \rightarrow 2SiCl_4 + 3H_2 + 2C$$

Trisilane

It is a bright mobile colourless liquid which inflames in air with a loud detonation.

$$Si_3H_8 + 3O_2 \rightarrow 3SiO_2 + 4H_2 \uparrow$$

It is decomposed by water

$$Si_3H_8 + 6H_2O \rightarrow 3SiO_2 + 10H_2 \uparrow$$

Tetra silane $S_{14}H_{10}$

It is a colourless liquid having similar chemical properties as that of trisilane.

Penta and Hexa Silane, Si_5H_{12} and Si_6H_{14}

These compounds could not be studied due to the difficulty in their isolation.

Silico Acetylene (Si_2H_2)

It is prepared by the action of hydrochloric acid on calcium silicide.

$$CaSi_2 + 2HCl \rightarrow CaCl_2 + Si_2H_2$$

$$nSi_2H_2 \rightarrow (Si_2H_2)$$

It is a yellow crystalline solid and is stable in air. It is nonexplosive and is not similar to acetylene at all except in its structure.

Silico Ethylene

It is obtained by the action of hydrochloric acid on calcium monosilicide.

$$Ca_2Si_2 + 4HCl \rightarrow CaCl_2 + Si_2H_4$$

Si_2H_4 exists in its polymeric form, *i.e.*, $(Si_2H_4)_n$.

Its structure is similar to ethylene.

Structure : The structures of silanes are similar to those of the corresponding saturated hydrocarbons. Silanes are more reactive than corresponding saturated hydrocarbons and are decomposed to silicon and hydrogen when heated.

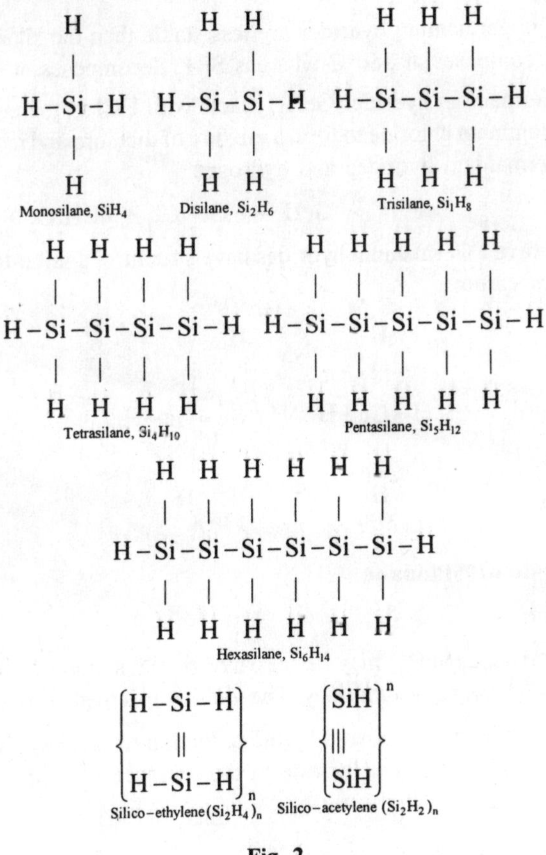

Fig. 2

GERMANIUM HYDRIDES

Preparation : Very little is known about the chemistry of germanium hydrides. Dennis, Corey and Moore first prepared a mixture of hydrides

(GeH_4, Ge_2H_6, Ge_3H_8,...) by the action of dilute HCl on magnesium germanide.

$$Mg_2Ge + 4HCl \rightarrow 2MgCl_2 + GeH_4$$

A pure sample of GeH_4 can be prepared by the action of sodium borohydride on an aqueous solution of germanium chloride. The yield is 74%. When calcium germanide is treated with aqueous hydrogen chloride, a non-volatile hydride $(GeH_4)_y$ is formed as a yellow solid.

Properties

(i) The germanium hydrides are less stable than the silanes. GeH_4 decomposes at 280°C whereas SiH_4 decomposes at 180°.

(ii) Germanium hydride (GeH_4) reacts with HCl in the presence of aluminum chloride to form a mixture of dichloro and monochloro germanium hydrides and hydrogen.

$$GeH_4 \rightarrow GeHCl_3 + GeH_2O_2 + H_2$$

Structure : Germanium hydrides have structures similar to those of hydrides of carbon.

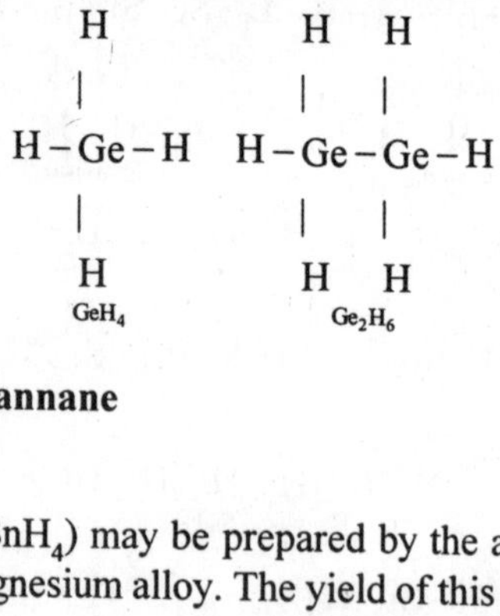

GeH_4 Ge_2H_6

Tin Hydride or Stannane

Preparation

(i) Stannane (SnH_4) may be prepared by the action of dilute acids on a tin-magnesium alloy. The yield of this method is very poor.

(ii) It can be best prepared by the reduction of tin (IV) chloride by lithium aluminium hydride in the presence of a little oxygen.

$$SnCl_4 + LiAlH_4 \xrightarrow{\text{ether}} SnH_4 + LiCl + AlCl_3$$

The yield of this reaction is 80% in the presence of oxygen. In the absence of oxygen the yield is only 20%.

(iii) When an aqueous solution of tin (II) chloride is treated with borohydride, stannane (SnH_4) is formed in 80% yield, some amount of very unstable distannane (Sn_2H_6) is also formed in this reaction.

Properties

(i) SnH_4 is a gas which boils at –52°C.

(ii) It decomposes to tin and hydrogen slowly at ordinary temperatures and rapidly at 100°C.

$$SnH_4 \rightarrow Sn + 2H_2$$

(iii) SnH_4 is a weak reducing agent. It only reduces silver nitrate and mercuric chloride solution.

(iv) The hydrogens of stannane (SnH_4) may be replaced by sodium.

$$SnH_4 + Na \rightarrow SnH_3Na + \frac{1}{2}H_2$$

$$SnH_4 + 2Na \rightarrow SnH_2Na_2 + H_2$$

(v) The poor stability of SnH_4 makes it difficult to use this hydride as an intermediate in the number of other compounds.

Structure : Stannane (SnH_4) possesses the following structure:

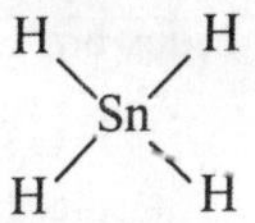

Lead Hydride or Plumbane PbH_4: For the first time, Paneth and Norring reported the formation of a volatile hydride of formula PbH_4.

(i) A small quantity of PbH_4 can be prepared by the electrolytic reduction of finely divided lead.

(ii) Small quantities may also be prepared by dissolving magnesium-lead alloy containing radioactive lead in dilute hydrochloric acid.

In the early work the amount of PbH_4 formed was so small that the existence of the hydride could only be demonstrated by using radioactive lead in its preparation as to obtain a gas, the radioacitivity of which was detected.

Properties

(i) It is colourless gas which decomposes rapidly at room temperature.

$$PbH_4 \rightarrow Pb + 2H_2$$

Lead hydride is less stable than stannane (SnH_4).

(ii) Very little is known about the properties of lead hydride.

Structure : The structure of lead hydride is similar to methane.

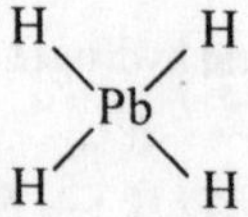

(d) **Metallic Hydrides or Interstitial Hydrides :** The transition elements and rare-earth metals combine with hydrogen to produce hydrides which are metallic in appearance but do not possess definite composition. These types of hydrides are known as metallic hydrides.

In these hydrides, the hydrogen atoms seem to occupy interstitial positions in the metallic lattice o and so form a solid solution. Because of this reason, metallic hydrides are also termed as interstitial hydrides. The metallic hydrides are related to other interstitial compounds such as borides, carbides and nitrides in the same way that the lattices of the parent metal undergo no change.

Some of the metallic hydrides are given below:

$LaH_{2.76}$, $Ce\ H_{2.69}$,

$PrH_{2.85}$, $TiH_{1.73}$,

$ZrH_{1.98}$, $TaH_{0.78}$,

$Pd\ H_{0.6}$ etc.

Apart form these simpler hydrides, transition metals form complex hydrides but their chemistry is of recent origin.

Preparation

(a) The metallic hydrides are obtained by occlusion or adsorption of hydrogen at moderate or high temperatures.

Palladium can adsorb 900 volumes of hydrogen. Similarly, other metals adsorb hydrogen forming metallic hydrides.

(b) Metallic hydrides may also be prepared by adsorption of hydrogen on a metal which is made cathode during electrolysis. The hydrogen liberated at cathode is adsorbed. For example,

(i) A palladium cathode may take up one thousand times its volume of hydrogen.

(ii) Electrolytically deposited iron may contain one hundred times its volume of hydrogen.

Properties : As the properties of metallic hydrides vary widely, it is therefore not possible to summarise the properties but it is possible to offer some generalisations.

(i) These hydrides are grey to black and have a metallic appearance.

(ii) They possess metallic properties such as hardness, lustre and electrical conductivity.

(iii) The quantity of hydrogen present in the metallic hydrides does not bear exact stoichiometric relation to the metal.

(iv) When metallic hydrides are heated at high temperatures, hydrogen is given out and therefore they behave as strong reducing agents. For example uranium hydride is a strong reducing agent, precipitating silver from silver nitrate solution and reducing cupric salts to copper.

The strong reducing properties of metallic hydrides suggest that hydrogen is present in the atomic state. It means that molecular hydrogen has undergone dissociation on entering the lattice.

(v) Formation of metallic hydrides may be accompanied by absorption or evolution of heat. For example :

(a) Copper silver, molybdenum, uranium, etc., form hydrides with the absorption of heat.

(b) Metals like titanium, thorium, zirconium, vanadium, palladium take up hydrogen exothermically.

Structure of Metallic Hydrides : The exact state in which the hydrogen gas exists in the metal lattice has been a subject of much investigation.

In many cases, adsorption of hydrogen results in no fundamental changes in the metal lattice, although, density and X-ray determination show the lattices to be expanded and perhaps slightly distorted. X-ray measurements indicate that at certain temperatures, the hydrogen atoms are accommodated in the interstices between the metal atoms in the crystal lattice. The hydrogen is given out on raising the temperature of the metallic hydrides.

Examples of Metallic Hydrides

(i) **Copper Hydride :** It is an example of metallic hydride which possesses exact stoichiometric composition (CuH). It is prepared by reduction of copper (II) solution with the sodium salt of hypophosphorus acid at about 70°C. Copper hydride forms red brown powder containing water of crystallisation. X-ray diffraction shows that copper atoms in CuH have been shown to be in a face-centred cubic arrangement as in metallic copper.

(ii) **Uranium Hydride and Uranium Deuteride :** These are also representative of this class of hydrides. Uranium metal reacts rapidly hydrogen and deuterium at 250-300°C to give a black powder. These reactions are reversible :

$$U + 3/2H_2 \rightleftarrows UH$$

$$U + 3/2D_2 \rightleftarrows UD.$$

Uranium hydride is extremely reactive and can inflame in air.

Uses of Hydrides

(i) **As Condensing Reagents :** Electrovalent hydrides are used as condensing agents in Claisen and aldol condensation.

(ii) **Source of Hydrogen :** Calcium hydride was used for generating hydrogen.

(iii) **As Reducing Agent :** As interstitial hydrides are capable of liberating hydrogen even on simple heating, they are very good reducing agents. The most important hydride which is used as a reducing agent is $LiAlH_4$.

(iv) **As a Fuel :** Diborane is being used in high energy fuels and propellants. Diborane is also used in welding torches.

(v) **General Uses :** Interstitial hydrides are used in metallurgy, in the manufacture of vacuum tubes, and catalytically active metals. Calcium hydride is used to determine traces of water in organic solvents.

BORAZINE

Introduction : The most important ring system of organic chemistry is the benzene ring either as a separate entity or in polynuclear hydrocarbons such as naphthalene, anthracence, and phenanthrene. Inorganic chemistry has two (at least) analogues of benzene: borazine and trimeric phosphonitrilic compounds.

Borazine, B_3N_3He, is of special interest because it is isoelectronic with benzene and some of its physical properties show some striking similarities with this (Table 1). Indeed, the earlier name borazole was proposed for borazine because of the resemblance to benzene (benzol) and it has been frequently referred to as inorganic benzene. The systematic name is s-triazatriborine, the compound is still known as borazine.

Table 1 : Comparison of the physical properties of Borazine and Benzene

Compound	*Mol wt.*	*M P. °C*	*B P. °C*	*Δ Hvap KJ mole^{-1}*	*Trouton's constant*
Borazine	80.6	–58	64.5	29.3	21.4
Benzene	78	+6	80.0	31.0	21.4

BORAZINE STRUCTURE

X-ray crystallography has shown that boron and nitrogen atoms in borazine are arranged alternately in a planar, hexagonal ring, in which the B–N bond length is 0.147 ± 0.007nm. Electron diffraction photographs of borazine and benzene are very similar and confirm that borazine has a benzene-like configuration. The B–N bond length determined by electron diffraction is 0.44 ± 0.002 nm. These facts are in accordance with the hexagonal ring structure.

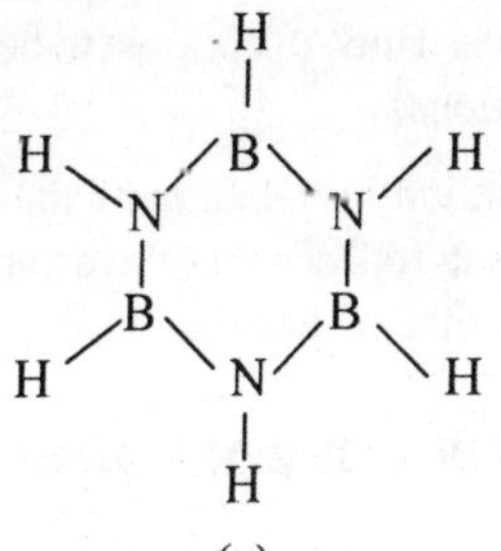

(a)

The bonding in borazine has been described in terms of resonance between the single-bonded triaminotriborine ring (a) and the Kekule structures (b) and (c) in which double bonds are present due to the participation of the lone pairs on nitrogen in the boron-nitrogen bonds.

This results in a formal negative charge on the boron and formal positive charge on the nitrogen atom. This resonance hybrid description is based on the similarities which exist between borazine and benzene. In borazine, the B–N bond length is 0.144 nm. which is very close to the value in benzene for C–C ad 1.42 nm.

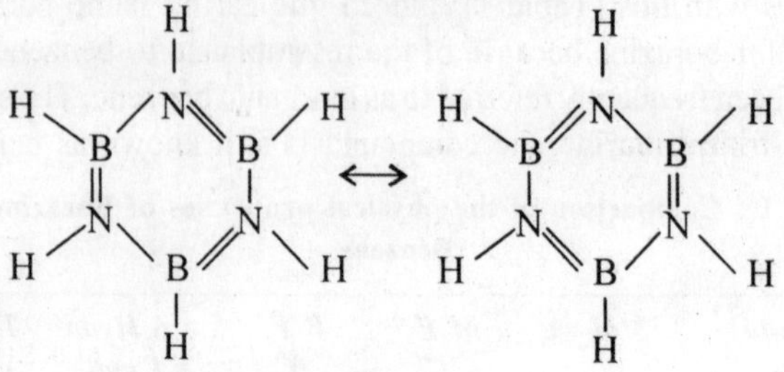

(b) (c)

Fig. 3

The molecular orbital description of borazine takes into consideration the formation of σ and π-bonds in the ring. Because, of the difference in electronegativity between boron and nitrogen, the π-cloud in borazine is lumpy with more electron density localised on the nitrogen than boron. This partial localisation weakens the π-bonding in the ring.

In addition, nitrogen retains some of its basicity and the boron some of the acidity. Polar species such as HCl can therefore attack the double bond between nitrogen and boron (see chemical reactions).

The different electronegativities of boron and nitrogen tend to stabilise bonding to boron by electropositive substituents and to nitrogen by electropositive substituents. Thus, in contrast to benzene, borazine readily undergoes addition reactions.

If a criterion of aromaticity is taken to be the extent of delocalisation of electrons in π-orbitals, it appears that the aromatic nature of borazine is significantly less than that of benzene.

PREPARATION OF BORAZINE METHODS

Method I

Borazine was originally prepared by Stock and Pohland in 1921 by the reaction of ammonia with diborane.

The adduct $(BH_3.NH_3)_2$ is first formed which is then decomposed by heating in a closed tube at 200°C. Cyclisation occurs to give borazine but yield is low because of the simultaneous formation of solid polymeric by-products.

$$3B_2H_6 + 6NH_3 \rightarrow 3(BH_3.NH_3)_2 \xrightarrow{200°C} 2B_3N_3H_6 + 12H_2$$

Method II

B-trichloroborazine, formed by the reaction between boron trichloride and ammonium chloride, can serve as an intermediate in the preparation of borazine. Small quantities are conveniently made by heating dry, powdered ammonium chloride in one end of a pyrex combustion tube at 170°C while a slow stream of boron trichloride vapour and nitrogen is passed over the powder.

B-trichlorobenzene sublimes as it is formed and condenses as needle-like crystals on the cool parts of the tube. Yields upto 35 percent have been reported. It is readily purified by vacuum sublimation. B-Trichlorobenzene is then reduced to borazine by sodium tetrahydroborate.

$$3NH_4Cl + 3BCl_3 \rightarrow 9HCl + Cl_3B_3N_3H_3$$

$$Cl_3B_3N_3H_3 + 3NaBH_4 \rightarrow B_3N_3H_6 + \frac{3}{2} B_2H_6 + 3NaCl$$

As the equation shows, $\frac{3}{2}$ moles of diborane are liberated for every mole of $B_3N_3H_3Cl_3$ used. This necessitates special precautions and the reaction is difficult to exploit on a large scale unless immediate use can he made of the diborane and suitable storage facilities are available.

An alternative procedure is to prevent the evolution of diborane by the addition of a tertiary aliphatic amine of relatively high boiling point. This reacts rapidly and completely with diborane to give the solid amine borane adduct.

Borazine is distilled in vacuous and purified by fractionation.

Method III : Borazine is also prepared by heating a mixture of lithium borohydride and ammonium chloride in vacuum at 230°C.

$$3NH_4Cl + 3LiBH_4 \xrightarrow{230°C} B_3N_3H_6 + 3LiCl + 9H_2$$

This reaction is the basis of a convenient laboratory method for borazine.

Lithium borohydride, dissolved in diglyme, is used as the reducing agent. The yield of this method is 80 percent.

PHYSICAL PROPERTIES OF BORAZINE

Borazine is a colourless liquid having boiling point 64.5°C. Borazine decomposes slowly on storage even at 80°C, and small quantities of a white solid are deposited over a period of several days. At room temperature, decomposition is much more rapid and hydrogen diborane and other volatile products are formed.

Borazine is isoelectronic which benzene. In physical properties, borazine is indeed a close analogue of benzene. The similarity of the physical properties of the alkyl substituted derivatives is more remarkable.

The ratio of the absolute boiling points of the substituted borazines to similarly substituted benzene derivatives is 0.93 ± 0.01. This similarity of physical properties has unfortunately been overemphasized and led to a description of borazine as inorganic benzene. However, the properties of borazine and benzene are quite different.

Chemical Properties

Main chemical properties of borazine are outlined below which are quite different from benzene.

1. Borazine readily undergoes addition reactions whereas benzene does not.

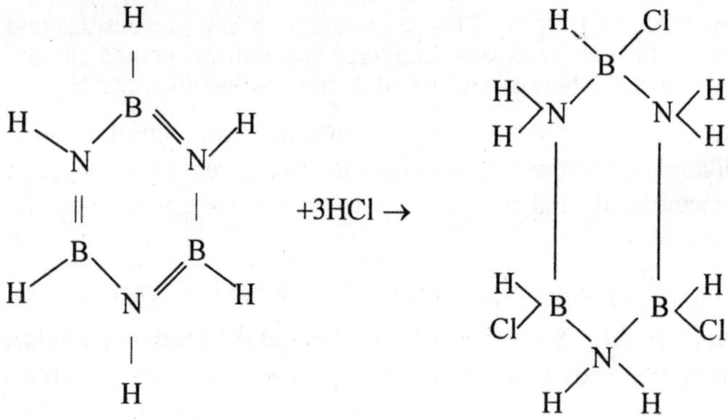

+HCl → No reaction

Fig. 4

The hydrogen chloride derivative of borazine, if heated to 50-100°C, loses three molecules of hydrogen to give B-trichloro-borazole.

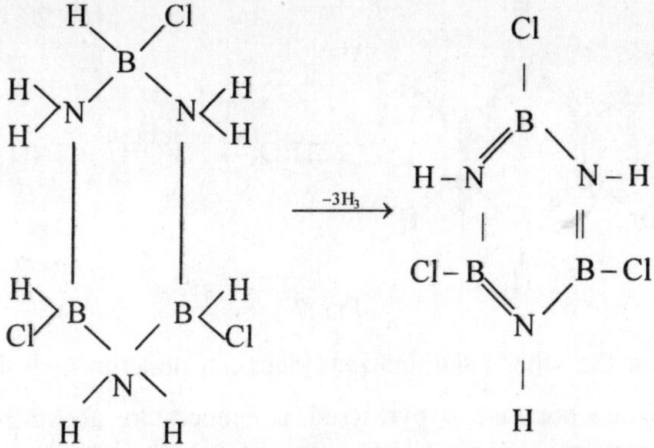

Fig. 5

2. Although, it was once believed that one mole of borazine forms an adduct with two moles of bromine and that this decomposes to B–2, 4-dibromobenzene as the main product, these reactions have not been repeated. It appears that borazine reacts with an excess of bromine to give an orange solid, B-hexabromo-cyclotriborazine, $B_3N_3H_6Br_6$, and this decomposes on heating to give B-tribromoborazine and HBr.

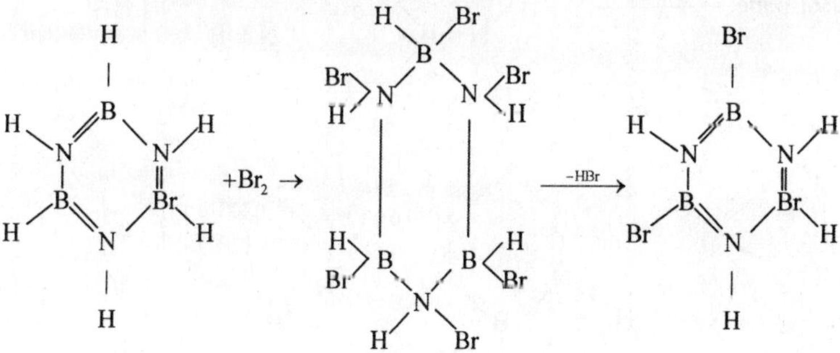

Fig. 6

On the other hand, benzene with bromine in the presence of ferric bromide yields monobromoben

$$+Br_2 \xrightarrow{FeBr_3} \overset{Br}{} + HBr$$

3. With water, borazine very slowly hydrolyses to hydrogen, boric acid and ammonia; this reaction is favoured by increase of temperature.

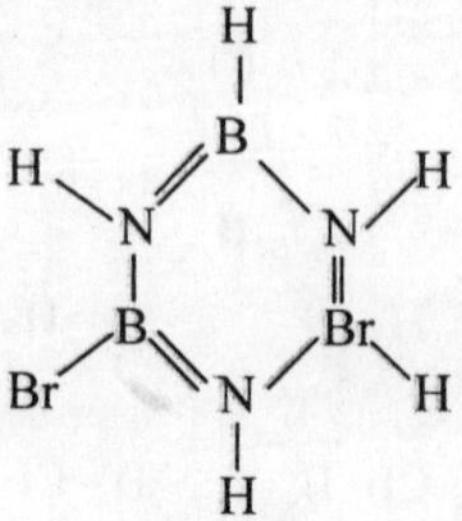

$$+3H_2O \rightarrow 2B(OH)_3 + 2NH_3 + 3H_2$$

Fig. 7

On the other hand, benzene does not undergo hydrolysis.

4. When borazine is pyrolysed at temperature above 340°C the products include $B_6N_6H_{10}$ (I) and $B_5N_5H_8$ (II). These are the boron-nitrogen analogous of diphenyl and naphthalene respectively. B-2, 4-diaminoborazine is also formed, presumably after a break down of some of the borazine molecules. The thermal decomposition of borazine is not completely analogous to that of benzene which gives biphenyl, terphenyl and triphenylene but no naphthalene.

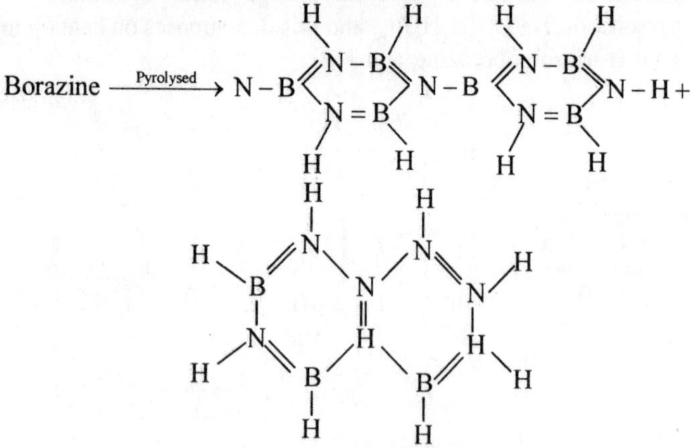

Fig. 8

5. Benzene may be hydrogenated to produce the saturated compound cyclohexane. Hydrogenation of borazine results in polymeric materials of indefinite composition.
6. Borazine forms an adduct with methanol in which borazine : methanol = I : 3. This pyrolyses with the elimination of hydrogen to give B-trimethoxyborazine.

7. Borazine reacts with phenylmagnesium bromide to give B- arylated compounds. The main products are B-mono-and B-diphenylborazine together with a small amount of B-triphenyl-borazine.
8. Borazine undergoes a strongly exothermic reaction with anniline to produce, triaminoborine (III).

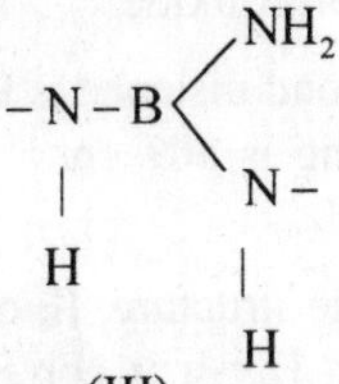

(III)

SUBSTITUTED BORAZINE

Main substituted borazines are discussed below

(a) **B-trimethyl borazine :** It is prepared by heating $(CH_3)_3B$ with ammonia at 320-340° at 20 atm for 2 hours. There are a number of stages to the reaction and the intermediate products can be isolated. However, the final product is B-trimethylborazine.

$$B(CH_3)_3 + NH_3 \rightarrow [(CH_3)_3B \leftarrow NH_3] \rightarrow (CH_3)_2B \overset{\pi}{-} NH_2 + CH_4$$

$$\downarrow$$

$$H_3B_3N_3(CH_3)_3 \leftarrow (CH_3)B \overset{2\pi}{-} NH + CH_4$$

At room temperature this compound (I) consists of colourless, volatile crystals. M. P. 31.8° B. P. 127°. It is stable in vacuum to 350°C. It is sensitive to moisture but is insoluble in water. It is soluble in many organic solvents. At 100°C water replaces the NH groups by oxygen and gives B-trimethyl boroxine (II).

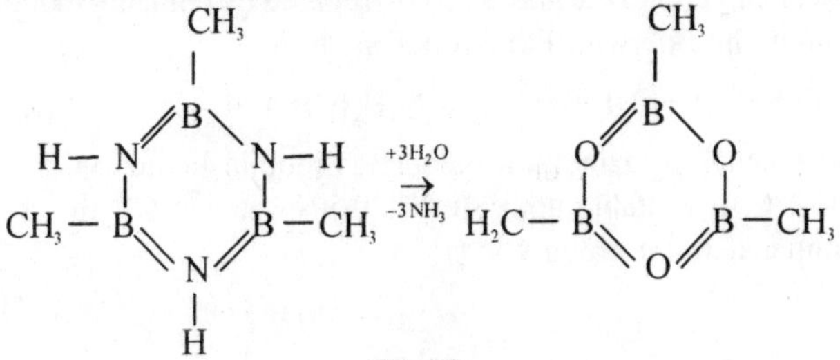

Fig. 9

(b) **Boroxine** : Isoelectronic with the borazine is boroxine, $H_3B_3O_3$(II). It is produced by the explosive oxidation of B_2H_6 or B_5H_9.

The structure has been determined and found to be perfectly planar. Boroxine is even less stable and presumably had less n delocalisation than borazine, decomposing at room temperature to diborane and boron oxide.

In boroxine the B–O bond distance is 0.138 nm. The characteristic Raman frequency of the ring is 807 cm^{-1}. Boroxine exhibits aromatic properties of benzene.

Note : By looking at the structure, II, one can immediately say that this structure is not correct. But it is correct.

The double bond between boron and oxygen is due to the participation of the lone pairs on oxygen. This result in a formal negative charge on the boron and a formal positive charge on the oxygen atom.

(c) **N-trimethyl Borazine** : It is obtained in 90 percent yield by heating a mixture of B_2H_6 and CH_3NH_2 in the correct proportions and at 180-200°C for 2 hours.

$$B_2H_6 + 2CH_3NH_2 \rightarrow 2[H_3B \leftarrow NH_2CH_3] \rightarrow 2[H_2B \overset{\pi}{-} NHCH_3] + 2H_2$$

$$\downarrow$$

$$2[HB \overset{2\pi}{-} NCH_3] + 2H_2$$

The final product polymerises to the trimer which is N-trimethyl borazine.

$$3\left(HB \overset{2\pi}{-} NCH_3\right) \rightarrow N-\text{trimethylborazine}$$

N-trimethyl borazine may also be prepared by reducing monomethyl ammonium chloride with lithium borohydride.

$$3CH_3NH_3Cl + 3\ LiBH_4 \rightarrow B_3N_3H_3(CH_3)_3 + 3LiCl + 9H_2$$

N-trimethyl borazine is a colourless mobile liquid, M. P. –90°C, B. P. 132°C. It is stable upto 300°C. With water at 60° the B–H link (III) is attacked and product is (IV).

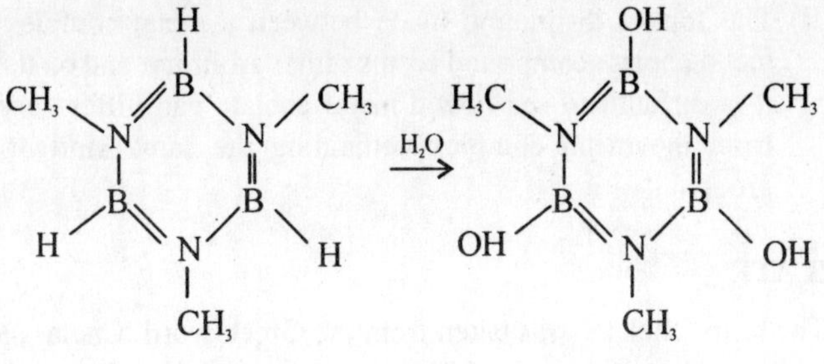

(III) (IV)

Fig. 10

CHELATES

Definition : The chelate compounds are defined as those complexes in which the donor atoms are attached to each other as well as to the metal so that the metal becomes part of a heterocyclic ring.

In other words the term 'chelate' was first of all introduced by Morgan and Drew to describible such cyclic structures which arise due to the combination of metallic atom with organic or inorganic molecules or ions. The rings of such compounds are termed as chelated rings and the phenomenon as the chelation. Simple examples of complex formation and chelation may be represented schematically in the following way.

A
↓
M + 4Ä → A →M ← A Complex
↓
A

A
|
M + 2Ä → Ä →M → A Metal chelate
|
A

where M represents the metal ion, A represents a complexing reagent and A–A represents a chelating reagent.

Limitations of Definition

(i) According to the above definition, there is no restriction on either the size of the ring or the number of such rings.

(ii) The formal distinction made between a metal complex and a metal chelate compound seems rather arbitrary, and on this basis it is difficult to see how a metal chelate can differ very much from the metal complex containing the same kind of donor group.

CHELATE

The term 'chelate' was taken from the Greek word 'Chela' meaning crab's claw. This analogy is obvious because both the claw and ligand may hold an object (*i.e.,* the metal ion) through more than one point of attachment.

CHELATE RING GROUPS

The rings may be closed by the formation of covalent linkages, or coordinate linkages or by a combination of both.

(a) **Covalent Linkage :** Covalent linkage is produced by the replacement of a proton in an organic group.

Functional groups of this type are sometimes called acidic groups because of the fact that hydrogen may be replaced from them. The most common examples of these are:

(i) — COOH (carboxyl group)

(ii) — SO_3H (sulphonic group)

(iii) — OH (enolic hydroxyl group)

(iv) = NOH (oxime).

(b) **Coordinate Linkage :** Coordinate linkages, without the replacement of hydrogen, are formed by the donation of an electron pair. The most common functional groups which contain donor atoms are:

(i) — NH_2 > NH and ≥ N (primary, secondary and tertiary amines)

(ii) = NOH (oxime)

(iii) — OH (alcoholic hydroxyl)

(iv) > C = 0 (carbonyl)

(v) —S— (Thioether).

CLASSIFICATION OF CHELATES

Dielh has classified chelates on the basis of:

(i) The total number of donor groups in the complexing agent *i.e.*, monodentate, bidentate, etc., and

(ii) The relative number of acidic and coordinating groups involved in ring formation.

Compounds which can link themselves with the central metal atom or ion through two bonds are called *two-toothed bidentate;* those capable of forming three bonds are known as *three-toothed* or *tridentate;* those forming four bonds are *quadridentate* and so on.

This classification takes into consideration both the number of bonds and types of bonds as given below:

1. **Bidentate groups :** Examples are oxalate, glycine and phenanthroline.

(a) **Two acidic groups :** Rings of this type are formed by the replacement of two hydrogen atoms by the metallic cation, *e.g.*, platinum ions react with oxalate ions to form platinum oxalate.

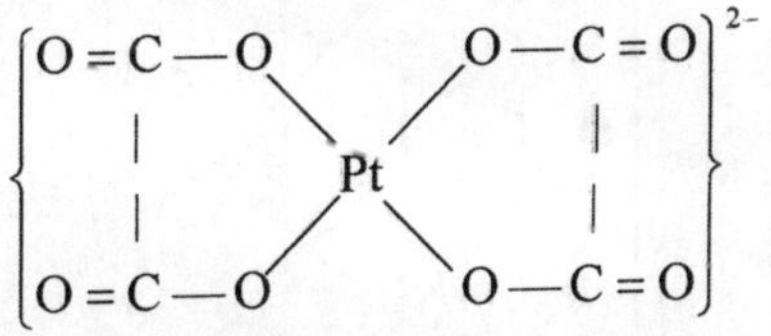

(b) **One acidic group and one coordinating group :** The cupric ion reacts with glycine to produce copper glycinate in which each glycine molecule is attached to the copper atom by one covalent bond and one coordinate bond.

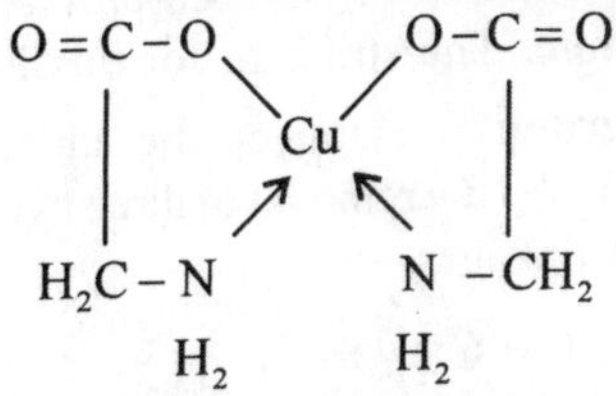

(c) **Two-co-ordinating groups :** Orthophenathroline is a heterocyclic organic compound having two ring of nitrogen atoms. This substance forms a complex ion with the ferrous ion in which three molecules of the orthophenanthroline are attached to one ferrous atom by two co-ordinating linkages. The formula of this chelate is given below :

Tridentate Groups

(a) **Two acidic groups and one co-ordinating group :** Aspartic acid combines with trivalent cobalt to form a water soluble tridentate compound. The structure given to this complex ion is

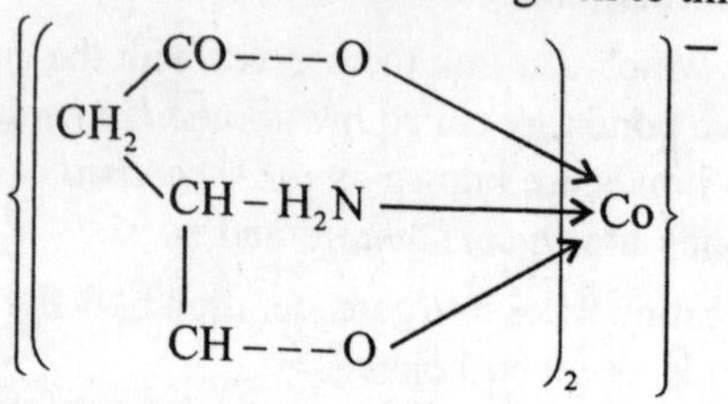

(b) **One acidic group and two co-ordinating groups :** The cobalt ion interacts with diaminopropionic acid to give a tridentate compound in which each molecule of the acid is attached to the cobalt atom by the two co-ordinating linkages and one covalent linkage. The formula for this chelate is

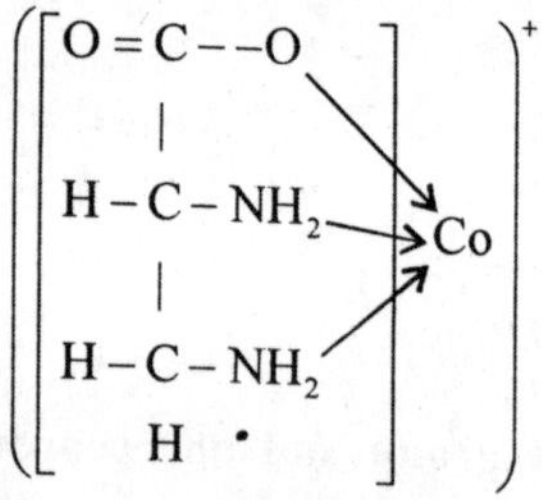

(c) **Three co-ordinating groups :** Triaminopropane combines with cobalt (III) to produce a chelate, in which one molecule of the triaminopropane occupies three co-ordination positions about the metallic atom. The structure for this is

(d) **Three acidic groups :** In tartar emetic, antimony atom is attached to the tartrate ion by the removal of three hydrogen atoms to form three covalent linkages.

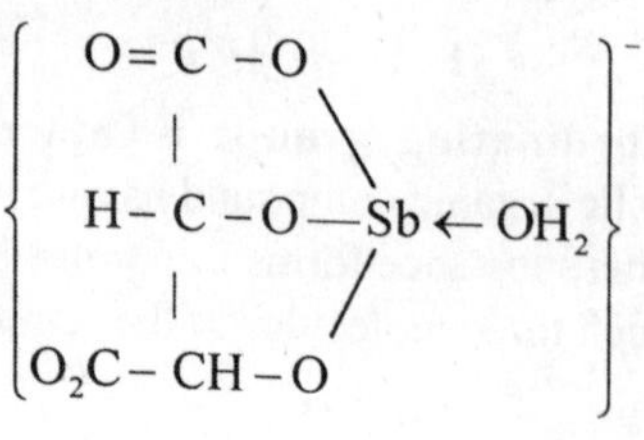

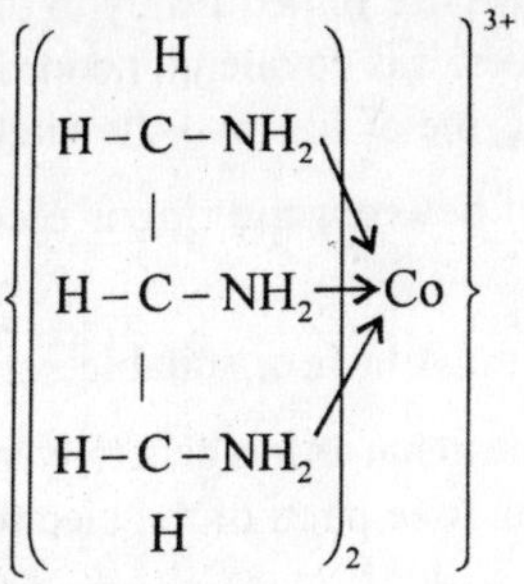

4. **Quadridentate and sexadentate groups :** Very few polydentate chelating agents beyond those capable of filling tridentate groupings are known

   (i) ***Two acidic groups and two co-ordinating groups :*** Ethylene diamine-acetic acid reacts with the cupric ion to produce a quadridentate grouping containing two covalent bonds and two coordinate linkages. The structure of this compound is:

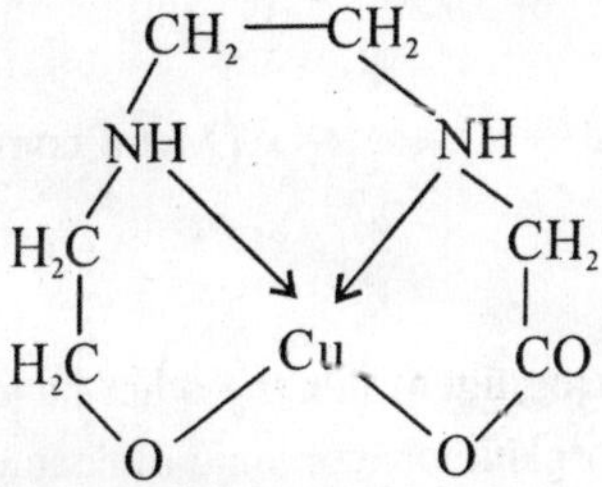

   (ii) ***Four co-ordinating groups :*** Tetrapyridyl reacts with certain bivalent transition metals, such as Fe^{2+}, Co^{2+}, Ni^{2+}, Cu^{2+} and Zn^{2+} to form a quadridentate compound in which the organic molecule is attached to the metal by four co-ordinating linkages. In the case of Fe^{2+}, the structure of the chelate.

CHARACTERISTICS OF CHELATES

1. They do not exhibit any of the properties of the salts. It is because that central metal atom has valency more than one. Chelates possess characteristics different from salts which are:

   (a) They have low melting points.

   (b) They are quite stable.

   (c) They diSSolv easily in organic solvents.

2. The donor groups are linked firmly to the metal ions present in the ring. However, it is covalent or coordinated covalent linkage which holds the metal ion into the chelating ring.
3. The presence of heterocyclic ring is essential for the formation of chelates.
4. The metal ion must have a suitable size.
5. The charge on the metal ions is generally large which is responsible for attracting the lone pairs of the electrons of the donor atoms.
6. In most of the chelates, the number of atoms forming the ring of the chelate is five or six but the former is more stable.

DISTINCTION BETWEEN METAL CHELATES AND METAL COMPLEXES

In a metal complex, the metal combines usually with an electron donor, not giving rise to cyclic rings.

$$\begin{array}{c} \mathrm{x} \\ \downarrow \\ \mathrm{M} + 4\ddot{\mathrm{x}} \rightarrow \mathrm{x} \rightarrow \mathrm{M} \leftarrow \mathrm{x}\ (\text{Metal complex}) \\ \downarrow \\ \mathrm{x} \end{array}$$

where $\ddot{\mathrm{x}}$ is a **complexing agent**. On the other hand in metal chelate the metal combines with organic or inorganic molecules or ions which may use two-atoms donor, giving rise to cyclic rings.

$$\begin{array}{c} \ddot{\mathrm{x}} \\ | \\ \ddot{\mathrm{x}} \end{array} + \mathrm{M} \rightarrow \begin{array}{c} \mathrm{x} \\ | \\ \mathrm{x} \end{array}\!\!\!\!\begin{array}{c} \diagdown \\ \\ \diagup \end{array}\!\mathrm{M}\ (\text{Metal Chelate})$$

where $\ddot{\mathrm{x}}—\ddot{\mathrm{x}}$ is **chelating agent**.

THE STABILITY OF METAL CHELATES

1. **Size of the chelate ring :** In general it has been reported by many workers that saturated compounds tend to form five membered structures while those ligands which contain double bonds tend to form six-membered rings.

Ley from conductometric studies of amino acid chelates concluded that chelates, with five and six membered rings are most stable. According to Schawarzenbach maximum chelate stabilisation is attained when a strainless rigid ring of a size, that can just accommodate the metal ion, is formed.

2. **Number of rings within the chelate :** It has been known for long that increase in the number of rings within the chelate results in greater stability. Quantitative studies have confirmed this. It is also found that the formation of fused rings around the metal tends to provide greater stability than the formation of single rings. For example, the chelate (I) having three interlocked ring system (fused rings) is much more stable than the chelate (II) involving comparable two ring system.

3. **Resonance effect :** The importance of resonance effects on the stability of metal chelates has been investigated by Calvin and Wilson (1945). They came to the conclusion that unhindered resonance of the chelate ring enhances the stability. On the other hand resonance in the attached benzene rings may interfere with resonance of the chelate ring and in such cases, stability of the chelate is lowered down.

4. **Steric effect :** This effect modifies the stability of the metal chelate to a great extent. Morgan found that acetyl-acetone did not give the usual coloured chelates with cupric or ferric ion when the 3-position was substituted by an isopropyl or secondary butyl group, but gave typical chelates when straight chain alkyl groups were present. This seemingly strange behaviour is due to steric effect of isopropyl and secondary butyl groups.

5. **Basic strength and chelate stability :** The more available the electron pair of the donor group the stronger is the covalent bond which it may form with a metal.

6. **Effect of substitution on the ligand :** The substitution of a group on the chelating agent may influence the tendency for chelation in one or two ways:

   (a) It may influence the acidity of the donor groups or may interfere with or enhance the resonance of the chelate rings.

   (b) The addition of groups on the ligand due to purely steric effects may prevent the ligand ions of molecules from acquiring the orientation about the central metal ion. (most favourable for chelation).

Substitution of an ethoxyl groups in place of a methyl groups of acetyl acetone greatly decreases the stability constant. This is due to the interference of the strong resonance of ethyl acetoacetate which prevents the carbonyl group from taking part fully in the resonance of the chelate ring.

7. **Effect of metal ion :** Transition metals usually form covalent bonds with electron donors. On the other hand, in the chelate compounds of alkaline earths and alkali metals, ions must form ionic bonds.

However, the stability of transition metal cheletes decreases with increasing basicity of the metal. Those transition metals which form the strongest covalent bonds with the ligand molecules produce stablest chelates.

The stability decreases with increasing basicity of the metal : Also metals which form the strongest homopolar bonds form the most stable chelates.

For chelates of the primary ionic type such as those of the alkaline earth and alkali metals the strength of chelation should increase with increasing ionic charge for both metal ion and donor and increase with decreasing ionic radius of the metal ion.

IMPORTANCE OF CHELATES

Although, the first chelate was recognized and described by Werner, the development and main importance of chelates has taken place almost entirely within recent years.

1. Among the earlier investigations Willstatter and H. Fischer, established the general structures of chlorophyll and hemin. Chlorophyll is a magnesium and hemin is an iron chelate.

2. **Analytical Chemistry**

   (a) **Qualitative Analysis :** Nickel, magnesium and aluminium in qualitative analysis are identified by the formation of stable, highly coloured chelates.

   Thus, dimethyl glyoxime, 8-hydroxy quinoline, o-phenanthroline, Fehling's solution and aluminon are used for the detection of many metal ions. For example, dimethylglyoxime reacts with Ni (II) to form an insoluble scarlet red precipitate. 8-Hydroxyquinoline is used for the determination and detection of over 30 elements such as Mg, Al, Zn, etc. The precipitates formed by this reagent with all these metals involve highly stable and characte-ristically coloured chelates.

(b) **Gravimetric Analysis :** Chelating agents are used in the gravimetric analysis of magnesium, nickel, etc., by their quantitative precipitation. The advantage of weighing the chelate is the high molecular weight of the precipitate and the low metal contents. This makes possible the determination of small amounts of metal with great accuracy.

(c) **Volumetric Analysis :** Chelating agents are frequently used as indicators in the titration of certain metal ions.

For example, ferroin, which is a chelate complex and contains $[Fe(phen)_3]^{2+}$ ion, is deep red in colour and acts as an indicator in oxidation-reduction titrations. In these titrations, the removal of the colour of the ferroin indicates that the end point has been reached. The ferroin is oxidised only after the reducing agent being titrated is completely oxidised.

3. **Water Softening :** Ion exchange resins have been produced which can successfully remove both cations and anions from aqueous solutions and the purity of water is nearly that of distilled water.

4. **Removal of Metal Ions :** The sequestering agents, which form water-soluble chelates, are used in aqueous solution for the effective removal of objectionable metal ions. These agents are also used in the removal of boiler scale during laundering.

5. **Elimination of Radioactive Metals :** Chelating agents such as ethylenediamine tetraacetic acid are used to speed up the elimination of harmful radioactive metals from the body. In using such agents, side effects, if any, should carefully be compensated.

For example, EDTA can be used to counteract the effect of lead poisoning but at the same time it may result in decalcification of bones and teeth because it (EDTA) forms a very stable calcium chelate. In order to prevent such decal-cification, EDTA is used in the form of $Na_2CaEDTA$.

6. **For Solvent Extraction :** Chelates which form soluble metal chelates with precipitates are used for solvent extraction.

   For example, the separation of Zr and Hf was a difficult analytic task until it was found that the chelates of these two metals with theonyl trifluoroacetone possess different solubilities in benzene. The distribution coefficient of these two metals chelates differ by a factor of 20, *i.e.,* zirconium has a greater tendency to be extracted. Therefore, it is now possible to effect a good separation of these two metals at present.

7. **Food Preservation :** Fruits, fruit juices and food stuffs are preserved with the help of chelating compounds. The addition of one hundredth of one percent of sequestering agents such as EDTA improves the qualities of the food on standing by catching the traces of metals (which are responsible for spoiling the food) by chelate formation.

DETECTION OF CHELATES

Most of the methods merely demonstrate complex formation. Thus, the, chelate nature of the complex ion is often infered by others, more or less indirect methods. Some of these methods are described below:

(a) **Absorption Spectra :** Many workers recognised that a characteristic colour is one of the most important and distinguishing features of chelate compounds, and used absorption spectra to distinguish between free metal ions and their chelates. Intensification of colour is usually considered an indication that chelate may be present.

   Simple colorimetric methods are used to determine the compositions of chelate compounds in solution. If the chelate is the only coloured substance present, the optical density of the solution is proportional to chelate concentration. Hence a plot of optical density versus composition of the solution would give a curve with a maximum at the composition corresponding to the formula of the chelate.

(b) **pH Effect :** All metal chelates may be considered as formed by the displacement of one or more usually weak acidic protons of the chelating agent by the metal ion. Thus, the addition of glycine to a solution of a copper salt causes a drop in the pH. This is due to the following equilibrium.

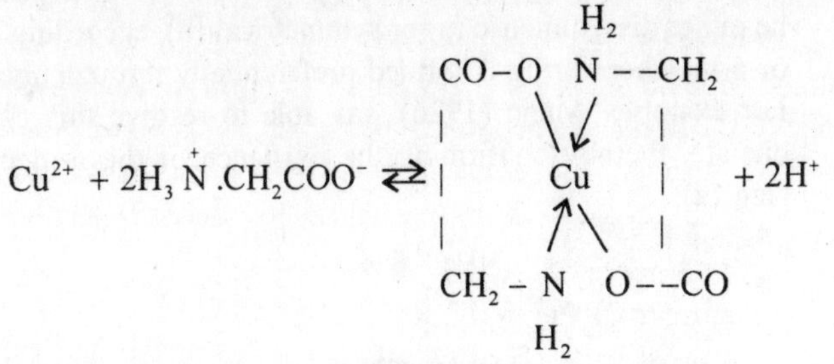

Fig. 11

Thus, the chelate formation of metal ions with amino acids and with the acid form of other chelating agents, results in pH drop and the greater the tendency for metals to combine with a given chelating agent, the greater the drop in pH.

The pH effect can be used as an indication of the extent of chelation.

(c) **Polarographic Measurements :** Valuable information about the composition of metal chelates and their behaviour in aqueous solution has been obtained from a study of their reduction at the dropping mercury electrode. The polarographic technique is based on the fact that the reduction potentials of metal ions are shifted to more negative values, as a result of chelate formation, and the correlation of this displacement of potential with both the concentration of the ligand and with solution, and pH provides an excellent method for determining the composition of metal chelate compounds.

(d) **X-ray Measurements :** The only direct proof for the existence of chelate ring is provided by X-ray crystal analysis. All other methods are indirect. The main difficulty lies in obtaining crystals possessing sufficient symmetry.

(c) **Optical Activity :** The resolution of chelate compounds into optically active forms has provided powerful evidence in the configuration of ring structures. For example, 1, 2, 3-triaminopropane can react with a metal so as to occupy only two coordination positions, the third amine group then being capable of salt formation.

The complex compound of this type formed with $PtCl_4$ will then he either dissymmetric (a) or symmetrical (b), according as a 5- or 6-membered ring is formed preferentially through chelation. For example, Mann (1926) was able to resolve this complex chelate, thereby confirming the existence of the 5-membered ring (a).

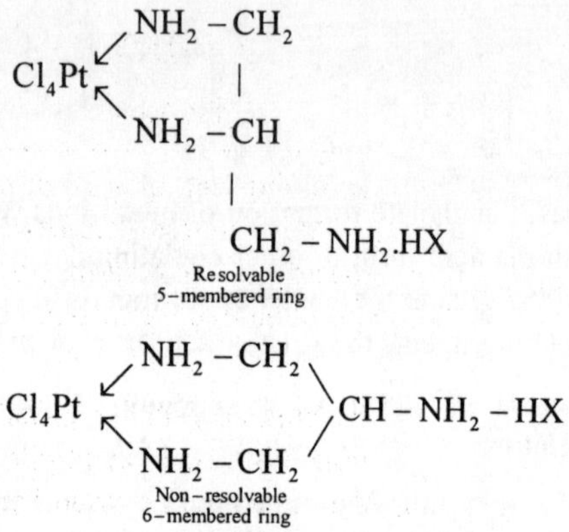

Fig. 12

This method cannot be applied to chelate rings in general since not all chelate structures are optically active.

(d) **Conductance Measurements :** Conductance measurements constitute probably the earliest method to study the formation of chelates. The disappearing of ions in complex formation may he followed by electric conductance measurements of the solution in which they are formed.

Chelate formation does not always result in a decrease in electrical conductance, since it is possible to increase the conductance of the medium by a suitable reaction in which highly conducting ions are produced in the course of reaction, *i.e.*, chelate formation.

For example, the addition of two molecules of ghycine to one molecule of aqueous copper acetate solution decreases the conductivity due to the formation of an uncharged chelated as

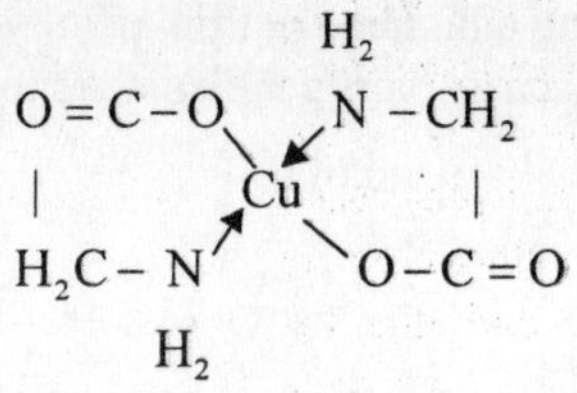

Fig. 13

Acetic acid produced in this reaction is a weak electrolyte, contributing little towards electrical conductivity.

Thus, this method is not reliable and gives only indication about complex formation and does not provide a demonstration of chelation.

ALKOXIDES

Introduction

The alkoxides are compounds in which a metal is attached to an alkyl group through oxygen and may be regarded as derivatives of alcohols, ROH, from which the hydroxylic hydrogen has been replaced by a metal or metalloid. The general formula of alkoxides is $M(OR)_n$ which M is metal of valence *n* and R is an alkyl group. Examples of alkoxides are :

$$NaOCH_3,\ Mg(OCH_3)_2,\ Al(OCH7)_3 \text{ and } Ti(OC_3H_7)_4$$

Alcohols reach with metals or their compounds under certain conditions and liberate hydrogen and alkoxides.

$$2ROH + 2Na \rightarrow 2RONa + 10_2$$

The order of ease of formation of an alkoxide with sodium or potassium has been found to be as follows :

Primary alcohol > Secondary alcohol > Tertiary alcohol

This can be explained as follows :

The Group –C–OH is able to undergo fission in two ways :

```
  |            |
  |            |
C–|–OH or, C – |O – – H
  |            |
```

As oxygen is having a higher electron affinity than either carbon or hydrogen, the shared electron pairs get displaced towards the oxygen *i.e.,* we have

$$\overset{\delta+}{C} : \rightarrow \overset{\delta-}{O} \leftarrow : \overset{\delta+}{H}$$

It is evident that larger the displacement of the shared pair towards the oxygen atom in the C–O bond, the larger the negative charge on the atom, and consequently the weaker would be the attraction of the oxygen atom for the shared pair of the O–H bond. In other words, the greater the polarity of the C–O bond, the smaller the polarity of the O–H bond.

As the polarity of a bond increases, the strength of that bond gets decreased. Hence, anything tending to increase the polarity of the C–O bond weakens this bond and strengthness the O–H bond, and *vice-versa.*

As alkyl groups are electron releasing, the larger number of alkyl groups, attached to the carbon atom of the C–OH group, the greater would be the polarity of the C—O bond. This may be represented as given below :

$$CH_3 \rightarrow CH_2 \rightarrow O - H \qquad \begin{matrix} CH_3 \searrow \\ CH_3 \nearrow \end{matrix} CH_3 \rightarrow \rightarrow O - H$$

Primary alcohol Ch_3 Secondary alcohol

$$\begin{matrix} CH_3 \searrow & \\ CH_3 \rightarrow C \rightarrow\rightarrow\rightarrow O - H \\ CH_3 \nearrow & \end{matrix}$$

Tertiary Alcohol

Because of this, the C–O bond becomes strongest in primary alcohols and weakest it tertiary and consequently the tendency of the C–OH group to break as C–OH becomes greatest in tertiary alcohols and lest in primary, conversely, the tendency of the C–OH group to break as C–O–:–H becomes greatest in primary alcohols and least in tertiary. This explains that reactions involving the breakage of the C–O bond will occur most readily with tertiary alcohols and least readily with primary ones. Qn the other hand reactions involving the breakage of the O–H bond will take place most readily with primary alcohols and least readily with tertiary ones.

Hence when sodium attacks on alcohol, hydrogen gets evolved as already indicated, and since this involves the breakage of the O–H bond, it can now be understood that why the order of ease of formation of alkoxides has been as follows :

Primary alcohol > Secondary alcohol > Tertiary alcohol.

The nature of alkoxides gets varied essentially from covalent volatile products in the case of electronegative elements like germanium and selenium to electrovalent ionic solids in the case of alkali or lanthanide metals.

CHARACTERISTICS OF ALKOXIDES

(i) *Metal oxides* M – O – C *bonds which are polarised due to strongly electronegative character of oxygen. The extent of this polarisation depends upon the electronegativity of the central atom, M. In the case of elements like germanium and selenium covalent volatile products are formed, whereas, in the case of alkali or lanthanide elements electrovalent solids are formed.*

(ii) *For derivatives of the same metal or metalloid the covalent character of metal oxygen bond* (M– O) *increases with the greater* (+ 1) *inductive effect of the alkyl group. Among, the isomeric butoxides of the elements the tertiary butoxide shows the highest covalent behaviour.*

(iii) *The polarity of the bond* $M^+ - O^-$ *may also be partially offset in the cases of some electrophilic elements which undergo covalence expansion by intermolecular co-ordination through the oxygen atoms of alkoxy group. The molecular association of this type appears to be due to steric factors as the ramification of the alkyl group increases.*

PREPARATION OF METAL ALKOXIDES

The method chosen for the preparation of an alkoxide of any element is generally determined by the electropositive character of element concerned.

(i) *The more electropositive metals react spontaneously with alcohols to form alkoxide.*

$$M + xROH \rightarrow M(OR)x + \frac{X}{2} H_2 \uparrow$$

More electropositive metals + Alcohol → Alkoxide + Hydrogen

(ii) *The metal chlorides of less elctropositive metals or metalloids (Si, Ti, Zr, Th etc.) undergo solvation and or partial solvolysis with alcohols, but the replacement of chloride is complete only in the presence of a base or al base acting as hydrogen chlorides acceptor such as the ethoxide ion, ammonia, pyridine, hydrogen chloride acceptor etc.*

$$TiCl_4 + 4ROH + 4NH_3 \rightarrow Ti(OR)_4 + 4NH_4Cl$$

$$TiCl_4 + 3EtOH \rightarrow 2HCl + TiCl_2(OEt)_2 . EtOH$$

Less electropositive metal chlorides + Alcohol → Alkoxide.

The various methods employed for the preparation of alkoxides are as follows.

1. **Action of Metal on Alcohols :** It is simplest method of preparing metal alkoxides.

$$M + xROH \rightarrow M(OR)_x + \frac{X}{2} H_2$$

This method is only applicable to more electropositive metals like Li, Na, K, Be, Mg and Al. The alkali metals are very reactive towards alcohols and accordingly the alkali metal alkoxides were the metal alkoxides to be prepared.

$$2Na + 2ROH \rightarrow 2NaOR + H_2$$

Beryllium magnesium, aluminium, etc. require catalysts for reactions with alcohols to proceed at a reasonable rate. The catalysts are generally iodine and mercuric chloride and their function is to disperse the insoluble oxide film which renders the metal inactive.

The nature of the alcohol is also an important factor for this method. For example, sodium reacts vigorously with methanol and ethanol whereas it reaction with isopropanol is considerably slower and with tertiary butanol the reaction is extremely slow.

An interesting example of this method is thallium ethoxide which can be prepared by holding the metal in air over the surface of alcohol.

$$2Tl + \frac{1}{2} O2 \rightarrow Tl_2O$$

$$Tl_2O + C_2H_5OH \rightarrow TlOC_2H_5 + Tl(OH)$$

$$Tl(OH) + C_2H_5OH \rightleftarrows TloC_2H_5 + H_2O$$

2. Action of Metal hydroxides on Alcohols : This method involves the reaction of metal hydroxides with and water is removed azeotropically with benzene. If water is not removed, alkoxide formed may undergo hydrolysis.

$$C_2H_5OH + NaOH \rightarrow NaOC_2H_5 + H_2O$$

Caldin *et al.* showed, by spectrophotometric method, that in ethanolic solutions having water, there is a formation of the elthoxide ion.

$$OH^- + C_2H_5OH \rightleftarrows H_2O + C_2H_5O^-$$

3. Methods involving the Use of Metal Chlorides : This method can be carried out in the absence or presence of ammonia depending upon the nature of metal.

(a) In the absence of ammonia : The alkoxides of polyvalent elements cannot be prepared by the above discussed method. However, they can be prepared by the reaction of anhydrous metal chlorides with alcohol.

$$SiCl_4 + 4C_2H_5OH \rightleftarrows Si(OH_2H_5)_4 + 4HCl$$

$$TiCl_4 + 3C_2H_5OH \rightarrow TiCl_2(OC_2H_5)_2.C_2H_5OH + 2HCl$$

$$2ZrCl_4 + 5C_2H_5OH \rightarrow ZrCl_2(OC_2H_5)_2.C_2H_5OH + 3HCl + ZrCl_3(OC_2H_5).C_2H_5OH$$

$$ThCl_4 + 4C_2H_5OH \rightarrow ThCl_4.4(C_2H_5OH)$$

From the above examples it is evident that the reaction between anhydrous chloride and alcohol results in the complete replacement of chlorine by alkoxy group only in a few cases. Therefore, this method must be carried out in the presence of ammonia or sodium hydroxide.

(b) In the present of ammonia : When the reaction of alcohol with metal chloride is incomplete, this must be carried out in the presence of ammonia or sodium hydroxide or pyridine whose presence allows the reversible reaction to proceed to completion by the removal of HCl liberated during the reaction as ammonium chloride or sodium chloride or pyridinium chloride.

$$MCl_n + nROH + nNH_3 \rightarrow M(OR)_n + nNH_4Cl$$

$$MCl_n + nRONa \rightarrow M(OR)_n + nNaCl$$

The alkoxides of zirconium, cerium, lanthanum, uranium, niobium, tantalum, iron, plutonium, antimony etc. have been prepared by this method. In all these cases, ammonium chloride or sodium or pyridinium chloride is precipitated whose separation can be facilitated by the addition of benzene in a few cases.

$$UCl_4 + 5ROH + NH_3 \rightarrow U(OR)_4 \xrightarrow{[O]} U(OR)_5 \text{ (Soluble in MeOH)}$$

$$UCl_4 + 5ROH + NH_3 \rightarrow U(OR)_5$$

$$LaCl_3 + 3LiOMe \rightarrow La(OMe)_3 \downarrow + 3LiCl$$

$$UCl_4 + 4LiOMe \rightarrow U(OMe)_4 \downarrow + 4LiCl$$

Zirconium tetra-alkoxide cannot be prepared by the reaction of ZrC14 with alcohol in the presence so sodium alkoxide due to the formation of stable double alkoxide such as $NaH[Zr(OC_2H_5)_6]$.

$$ZrCl_4 + C_2H_5OH + 5NaOC_2H_5 \rightarrow NaH[Zr(OC_2H_5)_6] + 4NaCl$$

When the above double alkoxide is treated with the equivalent amount of alcoholic hydrogen chloride, zirconium tetra-alkoxide is formed.

$$Nah[Z(OC_2H_5)_6] + HCl \rightarrow Zr(OC_2H_5)_4 + NaCl + 2C_2H_5OH$$

Zirconium tetra-alkoxide can be prepared directly by using anhydrous ammonia in place of sodium ethoxide.

$$ZrCl_4 + 4ROH + 4NH_3 \rightarrow Zr(OR)_4 + 4NH_4Cl$$

It has been reported recently that ammonia method is satisfactory for the preparation of pentaalkoxides of niobium and tantalum from their respective pentachlorides.

$$MCl_5 + 5ROH + 5NH_3 \rightarrow 5M(OR)_5 + 5NH_4Cl$$

The ammonia method is not successful in producing a chlorine free product because thorium alkoxides are stronger bases than ammonia in alcoholic solutions.

$$ThCl_4 + 4ROH + 4NH_3 \rightarrow Th(OR)_4 + 4NH_4Cl$$

$$Th(OR)_4 + NH_4Cl \rightarrow Th(OR)_3Cl + NH_3 + ROH$$

The sodium alkoxide method, however, results in the formation of pure thorium tetra-alkoxide.

$$ThCl_4 + 4ROH + 4NaOR \rightarrow Th(OR)_4 + 4NaCl$$

This method is successful due to the stronger basicity of the alkoxide ion as compared to the thorium alkoxide coupled with the smaller solubility of sodium chloride as compared to that of ammonium chloride.

4. From Double chlorides : The teteralkoxides of cerium and plutonium are obtained from the double chloride $(C_5H_5N)_2MCl_6$ because these two elements do not form stable tetra-chlorides.

$$(C_5H_5)_2MCl_6 + 6NH_3 + 4ROH \rightarrow M(OR)_4 + NH_4Cl + 2C_5H_5N$$

Zirconium tetra-alkoxides can be obtained from $(C_5H_5N)_2ZrCl_6$ which is regarded as a better starting material than $ZrCl_4$

$$(C_5H_5N)_2ZrCl_6 + 6NH_3 + 4ROH \rightarrow Zr(OR)_4 + 6NH_4Cl + 2C_5H_5N$$

It is observed that the use of zirconium tetrachloride often results in the formation of hydrolysed metal alkoxides.

5. By Reaction of Lithium Alkoxides with Metal Chlorides : The uranium (IV) tetra methoxide, uranyl methoxide and lanthanum trimethoxide are insoluble metal alkoxides and these can be prepared by the reaction of lithium methoxide with UCl_4, UO_2Cl_2 and $LaCl_3$ respectively.

$$UCl_4 + 4LiOCH_3 \rightarrow 4LiCl + U(OCH_3)_4$$

$$UO_2Cl_2 + 2LiOCH_3 \rightarrow 2LiCl + UO_2(OCH_3)_2$$

$$LaCl_3 + 3LiOCH_3 \rightarrow 3LiCl + La(OCH_3)_4$$

Lithium chloride, being soluble in methanol, is readily extracted out from the metal alkoxide.

6. Gilman's Method : Gilman and his coworkers prepared uranium pentaethoxide by the oxidation of uranium tetra-ethoxide by molecular oxygen which acts as an oxidant.

$$5U(Et)_4 + O_2 \rightarrow 4U(OEt)_5 + UO_2$$

The yield of this method is nearly 80%. However, the same workers achieved much better yield (90-95%) with bromine as an oxidant. For example uranium tetra ethoxide is first treated with bromine to form uranium (V) monobromide tetra-ethoxide which on subsequent treatment with sodium ehoxide yields uranium pentaethoxide.

$$U(OEt)_4 + \frac{1}{2} Br_2 \rightarrow UBr(OEt)_4$$

$$UBr(OEt)_4 + NaOEt \rightarrow U(OEt)_5 + NaBr$$

7. Reaction Between Metal Dialkyl amide and alcohol : Gilman et al., have prepared uranium (IV) alkoxides by treatment of terabis (diethylamido) uranium (IV) with appropriate alcohol. For example,

$$U(NEt_2) + 4ROH \rightarrow U(OR)_4 + 4Et_2NH$$

As the diethylamine ($HNet_2$) derivative is obtained from metal chloride, it means that the preparation of alkoxide is also done indirectly from the chloride.

This method is sometimes preferred over the direct method. For example, vanadium tetrachloride on treatment with certain alcohols results in the formation of $VCl_2(OR)_2.ROH$ which cannot be converted into $V(OR)_4$. However, vanadium tetra-alkoxides may be obtained as follows:

$$VCl_4 + 4LiNMe_2 \rightarrow V(NMe_2)_4 + 4LiCl$$

$$V(NMe_2)_4 + 4ROH \rightarrow V(OR)_4 + 4Me_2NH$$

8. By Alcoholysis Reaction : This method is based on the simple fact that there occurs and exchange of alkoxide group with alcohols in metal alkoxides to produce new alkoxides.

The reactions may be represented as follows :

$$M(OR)_x + R'OH \rightarrow \underset{H}{\overset{R'}{O}} \rightarrow \underset{\underset{O-R}{|}}{M(OR)_{x-1}} \rightarrow (R'O)M\,(OR)_{x-1} + ROH$$

$$U(OC_2H_5)_5 + 5ROH \rightarrow U(OR)_5 + 5C_2H_5OH$$

This reaction is reversible and is made to proceed in the forward direction by removal of ROH. The removal of ROH is only possible if ROH has a higher boiling point than R¢OH.

In this method of preparation speed, reversibility and absence of side reactions ply an important part. The reactions are generally carried out in inert solvents like benzene. Alkoxides of Be, B, Al, Ga, Si, Ti, Zr, Hf, V, Nb, Ta, Sb, Pr, Nd, Sn, Se, Te and Fe have been prepared by this method. Mixed alkoxides of a number of elements (Ti, Sn, Al, B, C, V, Nb, Ta and Se) have been successfully prepared in this manner.

9. Mehrotra's Process (Trans Esterification Reaction) : This method can be used for converting one alkoxide into another.

$$M(OR)_x + CH_3COOR' \rightleftarrows M(OR')_x + xCH_3COOR$$

This method possesses the advantage over the method (8) that separation of alkyl acetates is easier than the separation of alcohols. Further, this method has been found to be less susceptible to steric limitations than the method (8).

This method is particularly suitable for the preparation of tertiary butoxide from the isopropoxide derivatives, because the fractionation is very easy due to large difference in the boiling points of the esters.

Bradley and Thomas have recently extended the method for the preparation of trialkylsilyl acetate derivatives.

$$M(OR')_x + CH_3COOSiR_3 \rightarrow M(OSiR_3)_x + xCH_3COOR'$$

MISCELLANEOUS METHODS

(i) ***Meerwein*** *and co-workers have shown that many metal alkoxides are stable and that complexes of magnesium and calcium of the type are volatile and even distillable in vacuum.*

$$IM^{II} [Al(OC_2H_5)_4]_2$$

A similar uranium compound has been prepared by a group of scientists.

$$UCl_4 + 4Na[Al(OR)_4] \rightarrow U[Al(OR)_4]_4 + 4NaCl$$

(ii) ***Nesemeyanov*** *and* ***Nogina*** *prepared mixed alkoxides of titanium by the following reaction :*

$$TiCl_2(OR)_2 + R'OH + C_5H_5N \rightarrow TiCl(OR')(OR)_2 + C_5H_5N.HCl$$

$$TiCl(OR')(OR)_2 + NaOR'' \rightarrow Ti(OR'')(OR')(OR)_2 + NaCl$$

(ii) *Mixed alkoxides of several elements (Al, Ti, Zr, B, V, Nb and Sn, were prepared by* ***Mehrotra*** *and his coworkers by alcoholysis reaction in stoichiometric quantities of the reactions in the presence of benzene.*

$$Al\ (OPP^i_r)_3 + 2R^iOH \rightarrow Al\ (OPr^i)\ (OR^i)_2 + 2Pr^iOH$$

PROPERTIES

Physical Properties

1. Volatility Studies : The physical properties of metal alkoxides are mainly dependent on the *size* and *shape* of the *alkyl group,* and on the *valency, size, stereochemistry and co-ordination number* of the *metal.*

However, there occurs a gradual change in physical properties of alkoxides from non-volatile ionic alkoxides of alkali metals to the volatile monomeric covalent liquids of the higher transition elements through the polymeric covalent compounds of low volatility of the heavier multivalent element.

It has been observed that the physical properties of *covalent metal alkoxides* are decided by the resultant of two *opposing* tedencies, *i.e.*,

(i) *The tendency of the metal to expand its co-ordination number by intermolecular bond with oxygen, and*

(ii) *The steric effect of the alkyl groups.*

If the factor (i) is more powerful than the factor (ii), this results in the formation of relatively non-volatile alkoxides which may also be insoluble in organic solvents. If on the other hand, the factor (ii) predominates, volatile monomeric alkoxides are formed.

Bradley and *Wardlow* reported the effect of alkyl groups on the volatility of zirconium alkoxides. It is observed that the volatility decreases from ($Zr(OEt)_4$ to $ZrO(Pr^i)_4$ and finally to $ZrO(Bu^i)_4$ although there is an increase in their formula weights. The reason for this is that $Zr(OEt)_4$ and $Zr(OPr^i)_4$ are polymeric while $Zr(OBu^i)_4$ is monomeric. Thus, it reveals that the intermolecular bonding and hence volatility of the alkoxides is controlled by the steric effect of the alkyl group. Also, this study indicates that there is a direct correlation between the volatility of an alkoxide and its degree of polymerisation.

Martin and *Winter* have made a careful study of the variation in the degree of association of $Ti(OBu)_4$, $TiCl(OBu)_3$, $TiCl_2(OBu)_2$ and $TiCl(OBu)_3$ with concentration. They also reported that $TiCl(OBu)_3$, $TiCl_2(OBu)_2$ and $TiCl_3(OBu)$ are trimeric, dimeric and monomeric respectively in benzene with no noticeable concentration effect.

2. Thermal Stability : This study has not been made adequately. **Tishenko** (1990) reported the pyrolysis of various alkoxides of aluminium whose products include alcohols, ethers and olefins. *Meerwein* and *Geschke* reported the thermal decomposition of the ethoxides of tin, iron, cobalt, nickel and copper. For example,

$$Sn(OE)_4 \xrightarrow{\Delta} Sn(OEt)_2 + CH_3CHO + C_2H_5OH$$

$$Fe(OEt)_2 \xrightarrow{\Delta} 2Fe(OEt)_3 + CH_3CHO + C_2H_5OH$$

By the kinetic studies of zirconium alkoxides, *Bradley* and *Faktor* reported the following order for the zirconium alkoxides.

$$Zr(OEt)_4 >> Zr(OPr^i)_4 > Zr(Bu^i) > ZrO\ (Am^i)_4$$

In 1900 *tishchenko* showed that aluminium alkoxides on pyrolysis yield alcohols, ethers and olefins.

3. Magnetic Measurements : Magnetic susceptibility measurements were studied for many alkoxides. **Martin** and his coworkers have carried out careful measurements of magnetic moments and reflectance spectra of some di-and tri-valent transition metal alkoxides. The magnitude of moments at room temperature confirms that the compounds have high-spin configurations.

CHEMICAL PROPERTIES

1. Hydrolysis : Metal alkoxides undergo hydrolysis very readily. When these are treated with excess of water, metal hydroxides or hydrated oxides are generally precipitated. However, metal oxide alkoxides are obtained when hydrolysis is carried out very carefully. It is taken for molecule.

H   H
O: + $M(OR)_x$ →   O → $M(OR)_{x-1}$ → $M(OR)_{x-1}$+ROH
H   H |
                                          O – R

$M(OH)\ (OR)_{x-1} + M(OR)x \rightarrow (RO)_{x-1}\ MOM(OR)_{x-1} + ROH$

$2M(OH)(OR)_{x-1} \rightarrow (RO)_{x-1}\ MOM(OR)_{x-1} + H_2O$

2. Alcoholysis : Mehrotra *et al* studied the reactivity of alcohols on number of alkoxides of metals (Zr, Ti, V Sn. Ta etc). The order of reactivity reported is OMe > OEt OP^i > OBu^t. However, the mechanism of alcoholysis is analogous to the first stage of the hydrolysis reaction, *viz.*

                                R
R′OH + $M(Or)_x$ →   O: → $M(OR)_{x-1}$ (R′O)$M(OR)_{x-1}$ + ROH
                                H |
                                              O – R

with lower chain alcohols the alcoholysis is very rapid while with highly branched alcohols the interchange is slow and mixed alkoxides may be obtained.

3. Reactions with glycols : Alkoxides react with glycols in the 1 : 1 molar ratio to yield two moles of alcohol.

                    OH   O
M (OR)*n* + R   → $(OR)_{n-2}$M   R′ + 2ROH
                    OH   O

4. Reactions with anhydrous metal halides : Alkoxides of titanium, germanium and vanadium react with their chlorides and give chloride derivatives.

$$Ti(OR)_4 + 3TiCl_4 \rightarrow 4TiCl_3\ (OR)$$

5. Reactions with halogens and halides : Metal alkoxides react with halogens to form alkoxide derivatives. For example,

$$Ti(OCH_2R)_4 + X_2 \xrightarrow{X = Cl\ or\ Br} TiX_2(OCH_2R)_2 + 2RCH_2O$$

$$2RCH_2O \rightarrow RCH_2OH + RCHO$$

$$TiX_2(OCH_2R)_2 + RCH_2OH \rightarrow Tix_2\ (OH_2R)_2.RCH_2OH$$

$$2RCHO \rightarrow RCOOCH_2R$$

Metal alkoxides react with hydrogen halides readily whereby the alkoxide groups are replaced by halides.

$$M(OR)_x + yHX \rightarrow MXy(ORO)_{x-y} + yROH$$

Metal alkoxides also react with acetyl chloride giving different products.

$$Ti(OEt)_4 + CH_3COCl \rightarrow TiCl(OEt)_3 + CH_3COOEt \quad \textit{(Wardlaw)}$$

$$Ni(OEt)_4 + 3CH_3COCl \rightarrow TiCl_3(OEt).CH_3COOEt + 2CH_3COOEt$$

(Bradley)

$$Ti(OEt)_4 + 4CH_3COCl \rightarrow TiCl_4.CH_3COOEt + 3CH_3COOEt$$

$$Zr(OEt)_4 + CH_3COCl \rightarrow ZrCl(OEt)_3 + CH_3COOEt$$

(Bradley, Halim and Wardlaw)

$$Zr(OEt)_4 + 2CH_3COCl \rightarrow ZrCl_2.\ CH_3COOEt + CH_3COOEt$$

$$ZrCl_3(OEt).EtOH + 2CH_3COCl \rightarrow ZrCl_4.CH_3COOEt + HCl + CH_3COOEt$$

$$Zr_2(OBu^t)_4 + CH_3COCl \rightarrow ZrCl(OBu^t)_3 + CH_3COOBu^t.$$

6. Reaction with acetyl halides : When alkoxides are treated with acetyl halides, there occur the replacement of the alkoxy groups by halogens. For example :

$$Ti(OEt)_4 + CH_3COCl \rightarrow Ti(OEt)_3 + CH_3COOEt$$

Excess of acetyl chloride reacts with $Ti(OEt)_4$ as follows :

$$Ti(OEt)_4 + 3CH_3COCl \rightarrow TiCl_3(OEt).CH_3COOEt + CH_3COOEt$$

$$Ti(OEt)_4 + 4CH_3COCl \rightarrow TiCl_4.\ CH_3COOEt + 3CH_3COOEt$$

Zirconium ethoxide also reacts with acetyl chloride in a similar manner:

$$Zr(OEt)_4 + Ch_3COCl \rightarrow ZrCl(OEt)_3 + CH_3COOEt$$

$$Zr(OEt)_4 + Ch_3COCl \rightarrow ZrCl(OEt)_3\ COOEt + CH_3COOEt$$

7. Reaction with esters and anhydrides : The metal alkoxides undergo trans-esterification with carboxylic esters. For example.

$$CH_3COOR + M(OR')_x \rightarrow R'\text{–}O \quad \begin{matrix} OCCH_3 \\ O\text{–}R \\ M(OR')_{x-1} \end{matrix} \rightarrow CH_3COOR' + MOR(OR')_{x-1}$$

with carboxylic acid anhydrides, metal alkoxides form the metal carboxylates and alkyl acetates.

$$M(OR)_x + x\ (R'CO)_2O \rightleftarrows M(O_2CR')_x + xR'COOR$$

8. Reaction with carbonyl compounds : The well known reaction involving metal alkoxides and carbonyl compound is the *Meerwein. Ponndorf-Verley* reaction. For example,

$$R_2'CO + Al(OCHR_2)_3 \rightleftarrows \begin{matrix} R \\ C = O \\ R \end{matrix} + (R_2'CHO)\ (R_2CHO)_2Al$$

Meerwein Ponndorf Verley reaction involves catalytic reduction of ketones and aldehydes into secondary and primary alcohols respectively in the presence of metal alkoxides, *e.g.*,

$$R_2CO + R_2'CHOH \rightarrow R_2CHOH + R_2'CO$$

$$\begin{matrix} R \\ \\ R \end{matrix} C = O\left[(CH_2)_2\ CHO\right]_2 Al \rightarrow \left[\begin{matrix} R \\ R \end{matrix}\ CHO\right]_3 Al + \begin{matrix} CH_3 \\ CH_3 \end{matrix} C = O$$

Aluminium isopropoxide

$$\left\{\begin{matrix} R \\ R \end{matrix}\ CHO\right\}_3 Al \xrightarrow{\text{Dil, } H_2SO_4} \begin{matrix} R \\ \\ R \end{matrix} CHOH$$

Allyl alcohol

$$CH_2 = CH.CHO + 2H \xrightarrow{[(CH_3)_2CHO]_3Al} CH_2 = CH - CH_2OH$$

Sec. alcohol

$$CH_3 - CH = CH.CHO + 2H \xrightarrow{[(CH_3)_3CHO]_3Al} CH_2CH = CH - CH_2OH$$

Crotonaldehyde → Crotonyl alcohol

Mehrotra and his coworkers investigated reactions of salicyaldehyde with titanium and aluminium alkoxide.

9. Reactions with other hydroxy compound : *Mehrotra* and coworkers reported that soaps are obtained when alkoxides are treated with carboxylic acid. For example :

$$Al(OPr^i)_3 + 3R'COOH \rightarrow Al(COOR')_3 + 3Pr^iOH$$

When titanium and zirconium tetraalkoxides are made to react with excess of carboxylic acids, the three alkoxy groups of the alkoxides are replaced by carboxylic groups.

$$Ti(OR)_4 + 3R'COOH \rightarrow Ti(OR)\ (COOR')_3 + ROH$$

$$Zr(OPi)_4\ Pr^iOH + 3RCOOH \rightarrow Zr(OPr^i)\ (COOR)_3 + Pr^iOH$$

$$Al(OP^i)_3 + 3RCOOH \rightarrow Al(OOCR)_3 + 3Pr^iOH$$

Similarly zirconium compounds re prepared.

$$2Zr(OPr^i)_4 + 7RCOOH \rightarrow (RCOO)_3ZrO.Zr(OOCR)_3 + 7Pr^iOH + RCOOPr^i$$

Mehrotra and his coworkers studied the reaction of alkoxides of some elements (Sr, Ti, Al, Be etc.,) and carboxylic acid anyhydride.

$$M(OR)_x + x(R'CO)_2O \rightarrow M(OOR')_x + R'COOR$$

Alkoxides of some elements (Si, Ti, Al, Ge, Nd, Pr etc.) react with a-hydroxy carboxylic acid.

$$M(OR)_n + RCH(OH)COOH \rightarrow (RO)_{n-2}MOOC(O)CHR + 2ROH$$

10. Reactions with ethanolamines : Alkoxides of some elements (B, Al, Ti, Ge, Sn etc.) react with substituted alcohols (mono, di and tri, 2, hydroxy ethylamines), to form again an alkoxide. These reactions are similar to the alcohol intercharge reactions.

$$Ti(OR)_4 + HOC_2H_4NH_2 \rightarrow (RO)_3Ti(OC_2H_4NH_2) + ROH$$

$$Ge(OR)_4 + HOC_2H_4NH_2 \rightarrow (RO)_3Ge(OC_2O_4NH_2) + 2ROH$$

3

Phosphonitrilic Compounds or Phosphazenes

INTRODUCTION

Phosphonitrilic compounds or phosphazenes have attracted much attention these days because they form polymers with such properties which might be used in near future at high temperature.

Phosphonitrilic compounds are a group of compounds represented by the general formula $(NPX_2)_n$ *where* $X = F, Cl, Br, SCN, CH_3, C_6H_5$ *etc.*

The first systematic work on this subject dates form 1895 when Stokes commenced his pioneering research on the phosphonitrilic chlorides which are still the main starting materials for preparative work in this field. All these compounds are polymeric. These show similarities to aromatic compounds, although, the re-bonding is somewhat different.

Best known are the chlorides although mixed fluorine-chorine and fluorine-bromine derivatives are known. All the trimeric phosphonitirilic compounds have a planar P_3N_3 ring in which the hybrid P-N bonds are all equal in length whereas the tetramer is puckered and not planar as with the trimer. *Phosphonitrilic compounds are the compounds of phosphorus and nitrogen with a P-N* a *and a P-N π-bonds.*

Although, the π bonding has been somewhat different from that of aromatic compounds, the compounds resemble to some extent with aromatic compounds. These compounds have attracted attention because they form polymers with properties which might be useful at high temperatures. The materials' have been found to range from fluids possibly useful for heat transfer and lubrication to solids with some mechanical characteristics of organic polymers. Phosphonitrilic Chlorides. These are represented by the general formula $(NPCl_2)_x$ when x ranges form 3 to 7 and upwards. The monomar (x = 1) and dimer (x = 2), *i.e.*, $(NPCl_2)$ and $(NPCl_2)_2$ are not known. The following homologues are fully characterised:

Triphosphonitrilic chloride $(NPCl_2)_3$

Tretraphosphonitrilic chloride $(NPCl_2)_4$

Pentaphophonitrilic choloride $(NPCl_2)_5$

Hexaphosphonitrilic chloride $(NPCl_2)_6$

Heptaphosphonitrilic chloride $(NPCl_2)_7$

Polyphosphonitrilic choloride $(NPCl_2)_n$.

Out of the above, $(NPCl_2)_3$ and $(NPCl_2)_4$ have been fully studied.

Preparation

Method I: A mixture of phosphonitrilic chloride can be prepared by ammonolysis of phosphorus pentachloride.

$$3PCl_5 + 3NH_3 \longrightarrow (NPCl_2)_3 + 9HCl$$

$$4PCl_5 + 4NH_3 \longrightarrow (NPCl_2)_4 + 12HCl$$

Method II: Another method is by the reaction of phosphorus pentachloride with ammonium chloride in an inert solvent like tetrachloroethane.

A better procedure involves the heating of pentachloride with solid ammonium chloride at 145° to 160°.

$$xPCl_5 + xNH_4Cl \longrightarrow (NPCl_2)_x + 4xHCl$$

Mechanism: The formation of $(NPCl_2)_x$ may be explained as follows:

(i) *Ammonium chloride reacts with* PCl_5 *yielding* NH_4PCl_6. *ammonium hexachlorophosphate.*

(ii) NH_4PCl_6 *is unstable and decomposes to yield* $HN.PCl_3$ *and* HCl.

$$NH_4PCl_6 \longrightarrow HN.PCl_3 + 3HCl$$

(iii) *Now* $HN.PCl_3$ *undergoes condnsation to form* $(NPCl_2)_x$ *with loss of hydrogen chloride.*

$$xHN.PCl_3 \longrightarrow (NPCl_2)_x + xHCl$$

In the above method, a mixture of phosphonitrilic halides is obtained. Separation of the reaction products is affected by distilling off the solvent and treating the residue (which is a mixture of oil and crystalline material) with petroleum ether. The soluble part varying form 60-90 percent of the whole, is a mixture of cyclic polymers $(NPCl_2)x$, and the insoluble part consists of mixture of linear polymers $(NPCl_2)x. PCl_5$. The cyclic products from the soluble part are separated by distilling off the trimer at 124°C at a pressure of 10 mm. and the tetramer at 185°C at a pressure of 10 mm.

Other members of the series upto $(NPCl_2)_8$ may be separated by solvent extraction, sublimation and crystallisation, leaving a small inseparable residue containing higher members of the series with a mean molecular composition of about $(NPCl_2)_{12}$. Proof that only ring polymers are present in this material comes from the ^{31}P n.m.r. spectrum which has a single peak that all the' phosphorus atoms are in the same environment.

Method III: Another method is by heating $(P_3N_5)x$ in chlorine at 700°C.

$$(P_3N_5)x + xCl_2 \longrightarrow (NPCl_2)x + \text{other products.}$$

Method IV: S_4N_4 reacts with $SOCl_2$ in PCl_3 to give only the $(NPCl_2)_3$.

$$S_4N_4 + 6SOCl_2 + (12PCl_3 \longrightarrow 4(NPCl_2)_3 + 10S + 3O_2 + 12Cl_2$$

Isolation: The trimer and tetramer may be obtained by distillation of the mixture obtained in the above mixture, but only at the expense of polymerising much of the metreial.

Properties

The halogen atom in the cyclic phosphonitrilic chlorides undergo many of the reactions normally associated with the P-Cl bond, though sometimes less readily so that complete replacement of halogen by other groups may be difficult to affect. Most work has been done with $(NPCl_2)_3$ and $(NPCl_2)_4$. However, some of the properties are:

(i) ***Hydrolysis:*** *Chlorine atoms in* $(NPCl_2)x$ *are fairly reactive and on hydrolysis some or all the chlorine atoms can be replaced by hydroxyl groups.*

Hydrolysis of trimer is slow. However, partial or complete hydrolysis can be achieved by the use of aqueous ether.

(ii) $(NPCl_2)_3$ *melts at* 114°C, *boils at* 256°C *at* 1 *atm. pressure and is readily soluble in benzene, ether and carbon tetrachloride.*

$(NPCl_2)_4$ *melts* 123.5°C, *boils at* 328.5°C *and has a lower solubility in benzene, ether and carbon tetrachloride than the* $(NPCl_3)_3$.

(iii) $(NPCl_2)_3$ *and* $(NPCl_2)_4$ *are irritating and somewhat toxic also.*

(iv) $(NPCl_2)_3$ *or* $(NPCl_2)_4$ *on heating to* 250°C *polymerises to an elastic product like polyphosphonitrilic chloride or to a mixture of chlorides of high molecular weight of at least* 20000. *When the temperature rises above* 350°, *depolymerisation occurs. xygen is necessary for the reaction to occur and it may play a part both in ring cleavage and a chain-termination. Polymerisation is accompanied by decreased solubility in organic solvents and decreased reactivity.*

Trimer in acid solution is hydrolysed to orthophosphoric acid and ammonia. Hydrolysis of tetramer by boiling water is rapid and the product obtained is fairly stable. In the presence of silver ions, the acids from the trimer and tetramer give hexa and octa-silver salts. In case hydrolysis is done in acid solution, both the trimer and the tetramer give phosphoric acid and ammonia.

(v) ***Reaction with ammonia***: *Trimer reacts with ammonia to give products such as* $N_3P_3Cl_4(NH_2)_3$. *If the action is allowed to continue, all the chlorine atoms are replaced by amino groups giving* $N_3P_3(NH_2)_3$ *which loses* 3 *molecules of ammonia forming phospham.*

Phospham when heated give $(P_3N_5)x$ and ammonia. Tetramer undergoes the same reaction with ammonia.

(vi) ***Friedel-Craft's reaction.*** *Benzene reacts with* $N_3P_3Cl_6$ *in the presence of aluminium chloride (Friedel-Craft's reaction) to form diphenyl derivative.*

(vii) ***Reaction with amines***: *Reaction with primary and secondary amines gives a wide range of partially and fully substituted derivatives, viz.* $[NP(NH.C_6H_5)_2]_3$ *and* $[NPCl(NMe_2)]_3$.

(viii) ***Reaction with phenyl magnesium bromide***: *Reaction of phenyl magnesium bromide with the phosphonitrilic chlorides at about* 115° *also gives partially or completely phenylated product.*

Partially or completely phenylated product is also formed by reaction with benzene in the presence a Friedel-Crafts catalyst such as $AlCl_3$, *e.g.*,

$$(NPCl_2)_3 + 6C_6H_6 \xrightarrow{AlCl_3} [NP(C_6H_5)_2]_3 + 6HCl$$

(ix) **Reaction with alcohols and alkoxides:** *When treated with these reagents the chlorine atoms may be replaced by alkoxy groups. The fully substituted product* $[NP(OC_2H_3)_2]_3$ *isomerises at* 200°C.

(x) **Reaction with postassium thiocyanate:** *Potassium thiocyanate in acetone is able to convert* $(NPCl_2)_3$ to $[NP(NCS)_2]_3$

$$(NPCl_2)_3 + 6KCNS \longrightarrow [NP(CNS)_2]_3 + 6KCl$$

(xi) **Reaction with lead fluoride:** *When the trimer is heated with* PbF_2 *some of the chlorine atoms get replaced by fluorine along with the formation of tetramer.*

$$(NPCl_2)_3 + PbF_2 \longrightarrow (NPClF)_3 + N_4P_4Cl_2F_6$$

(xii) *Basic properties of phosphonitrilic chloride : The lone pairs of electrons on the nitrogen atom of phosphonitrilic derivatives are potential sites for attack by Lewis acids. The properties of phosphonitrilic halides are, however, very weak. Trimeric chloride forms* 1 : 1 *and* 1 : 2 *adducts with* $HClO_4$ *in which it is likely that protonation of nitrogen occurs. Attack is also probable on nitrogen in* $P_3N_3Cl_6.3SO_3$.

$$N_3P_3Cl_6 + HClO_4 \longrightarrow N_3P_3Cl_6.HClO_4$$
$$N_3P_3Cl_5 + 2HClO_4 \longrightarrow N_3P_3Cl_6.2HClO_4$$
$$N_3P_3Cl_6 + 3SO_3 \longrightarrow N_3P_3Cl_6.3SO_3$$

With $AlCl_3$, trimer forms adduct $N_3P_3Cl_6.2AlCl_3$ in which it is believed that there is a donation of :chloride ions to form $[AlCl_4]^-$.

$$N_3P_3Cl_6 + 2AlCl_3 \longrightarrow N_3P_3Cl_6.2AlCl_3$$

(xiii) *Lead fluoride fluorinates the trimer giving ultimately* $N_3P_3F_6$. *Some of trimer is converted into derivative of the tetramer such as* $N_4P_4Cl_4F_6$ *and* $N_4P_4Cl_2F_6$.

$$N_3P_3Cl_6 \xrightarrow{PbF_2} N_4P_4ClF_4 + N_4P_4Cl_2F_6$$

Also, tetramer adds on perchloric acid giving $(PNCl_2)_4.2HClO_4$.

(xiv) *They react smoothly with alcohols and alkoxides.*

$$(NPCl_2)_3 + 6NaOC_2H_5 \longrightarrow [NP(OC_2H_5)_2]_3 + 6NaCl$$

Structure and Bonding

X-ray examination reveals that the trimer and tetramer are having the cyclic structures in which the six- and eight-membered rings are composed of alternate nitrogen and phosphorus atoms.

Thus these substances may be regarded as the phosphorus-nitrogen analogues of benzene and cyclo-octatetraene. But as all the nitrogen-phosphorus distances are the same (1.60Å in-the trimer and 1.68Å in the tetramer), each formula must be considered as representing one canonical extreme of a mesomeric hybrid. It is to be noted that the P–N bond distance in these compounds is shorter than the P-N bonds. (1.78Å) in the phosphoramidate ion, $NH_6PO_3^{2-}$.

(a) Structure of triphosphonitrilic chloride: Most of the cyclic phosphonitrilic derivatives are crystalline solids and X-ray analysis has revealed interesting variation in the ring shape and dimensions. Tine trimeric has almost planar ring (Fig. 26.5). There is an approximately tetrahedral distribution of alencies round phosphorus corresponding with sp^3 hybridisation, with the result that the chlorine atoms. The each pair lie on opposite side of the rings. All the N–P bonds in the ring are also of equal length. It is 59 + 0.02 Å.

In the above structure it is evident that two chlorine atoms are attached to each phosphorus atom. This is confirmed by two chemical experiments such as:

(i) *Benzene reacts with* $N_3P_3Cl_8$ *in the presence of* $AlCl_3$, *thus*:

The product (III) is hydrolysed by water to $(C_6H_5)_2$ PO.OH. Hence both the chlorine atoms originally have been attached to the same phosphorus atom.

(ii) *Phenyl magnesium bromide at* 115° *also reacts with* $N_3P_3Cl_6$ *to produce diphenyl derivatives.*

When molecular orbital theory is applied to the structure of $N_3P_3Cl_6$, it is observed that there are normal loclised σ bonds through the ring formed by overlap of sp^3 hybrid orbitals of phosphorus with sp^2 orbital of nitrogen. In addition, each nitrogen atom has an electron in a p_z orbital and each phophorus atom in a d-orbital. These orbitals combine to give delocalised $p\pi$ - $d\pi$ orbitals which extend over the whole ring structure. The presence of negative chlorine atoms on phophorus makes the diffuse phosphorus d-orbitals more compact and favours their overlap.

As shown below, resonance structures can be drawn analogous to those for benzene indicating; aromaticity in the ring.

The resonance in the trimer is justified by the fact that all the N–P distances are the same (1.67 which are less than that expected for a P–N single bonds (1.78 Å) as in sodium phosphoramidate. However various reasons are there which point out that the double bond character is not very considerable and hence the extent of delocalisation of the π electrons is much less significant than in the benzene molecule.

Structure of Tetraphosphonitrilic Chloride, and Other Higher Chlorides

The tetrameric chloride $(NPCl_2)_4$ has a remarkable puckered ring structure of alternate phosphorus and nitrogen atoms, with two chlorines on each phosphorus atom. Its structure apparently involves Kekule resonance as well.

The chair and boat forms have been isolated for $(NPCl_2)_4$. It is probable that such ring structures may be written for penta, hexa and heptaphosphonitrilic chlorides. $(NPCl_2)_3$ is flat and $(NPCl_2)_4$ exists in the chair and boat conformation.

Phosphonitrilic bromides: The bromo compounds can be prepared in similar manners as being done for chloro derivatives. These compounds have the same structures.

Phosphonitrilic alkyls: These may also be prepared exactly in the same way. For example, Me_2PCl_3 reacts with NR_2Cl to give a mixture of products from which $(NPMe_2)_3$ and $(NPMe_2)_4$ have been isolated.

Phosphonitrilic fluorides: These are not prepared from PF_5 and NH_4F presumably because the, NH_4F acts as fluorinating agent and thus the sole product has been the stable hexafluorophosphate NH_4PF_6.

Thus the fluoro-analogues of phosphonitrilic chlorides can be prepared by the action of potassium fluorosul-ohinate (KSO_2F) on $(PNCl_2)_3$ or 4 at 120°.

$$(NPCl_2)_3 + 6KSO_2F \longrightarrow (NPF_2)_3 + 6KCl + 6SO_2$$

$$(NPCl_2)_4 + 8KSO_2F \longrightarrow (NPF_2)_4 + 8KCl + 8SO_2$$

The fluorine compounds have been much more volatile than the corresponding chlorides and there are easier to separate, *e.g.*, $(NPF_2)_3$ and $(NPF_2)_4$ have b.p. 51° and 246.7°C respectively. $(NPF_2)_3$ like $(NPCl_2)_3$ and $(NPF_2)_4$ unlike $(NPO_2)_4$ are having planar structures.

The fluoro trimer is insoluble in polar liquids, even in concentrated sulphuric acid. Both the trimer and tetramer are stable upto 300°C, but above this temperature they give colourless liquid polymers. The trimer derivative [$(NPF_2)_3$] has a planar ring whereas the compound $(NPF_2)_4$ is planar but not in puckered as in $(NPCl_2)_4$.

Phosphonitrilic chlorofluoride: A chlorofluoro-polymer $(NPFCl)_4$ is prepared by heating the trimer $(NPCl_2)_3$ with PbF_2.

$$(NPCl_2)_3 + PbF_2 \longrightarrow (NPFCl)_4 + PbCl_2 + \text{Other chloro-fluoro compounds.}$$

All the possible trimeric and tetrameric phosphonitrilic chlorofluorides have now been obtained. The structures of these compounds are not certain.

(NP Phenyl Cl_4) and [NP(NB Phenyl)$_2$]$_4$: The compounds can be prepared by treating with phenyl magnesium bromide and aniline respectively.

(NP Phenyl Br)$_4$ and (NP Phenyl Br)$_4$: These compounds can be prepared by the method:

$$\text{P Phenyl } Br_2 + NH_4Br + Br_2 \longrightarrow (\text{NP Phenyl Br})_3 + (\text{NP Phenyl Br})_4$$

+ oily polymers. However, the trimer shows geometrical isomerism, *i.e.,* cis and trans isomers.

Linear phosphonitrilic halides or Inorganic rubber: When phosphorus pentachloride and ammonium chloride are heated together in refluxing s-tetrachloroethane, that protion of the product which is insoluble in petroleu m ether has the approximate composition $(NPCl_2)_nPCl_5$. where π may be as small as 10 – 15. The possible structure of this compound is.

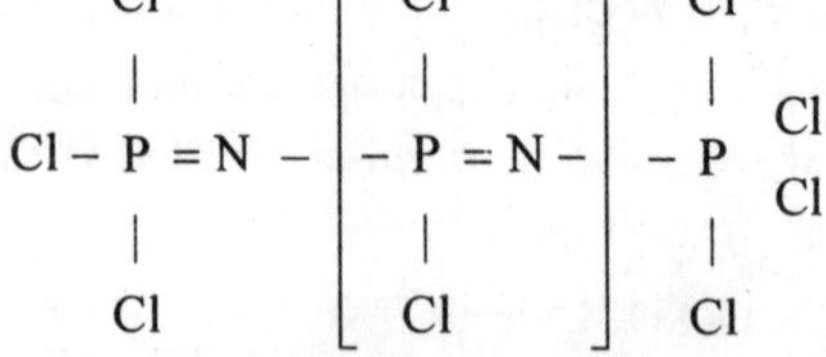

Fig. 1

This structure is consistent with the ^{31}P n. m. r. spectrum which shows two peaks, neither of which is due to free PCl_5. A very similar material is formed when $(PNCNI)_2$ is heated in a sealed tube at 350° with PCl_5.

When a cyclic halide such as $(NPCl_3)_2$ is heated to 250–350° in a solvent such as carbon tetrachloride with excess of oxygen, linear polymers with molecular weights in excess of 20,000 result. Oxygen necessary for the reaction to occur.

The materials are rubber- like and when stretched, give an X–rays diffraction pattern of the same type as natural rubber. They are believed to be made up of long spirit chains. Like rubber they consist of zig-zag long chains as shown below:

The above structure is end-stopped with the elements of PCl_5 at one end and PCl_4 at the other. Measurements of the modulus of elasticity of the rubber–like materials have indicated molecular weights in the range of 80,000 thereby indicating the presence of some 1000 atoms in the elastic chain ...–N–P–N–P–...

One of the drawbacks of inorganic rubber is its susceptibility to hydrolytic attack. Further if it is stored in air for several months, inorganic rubber becomes brittle due to hydrolytic condensation by the atomspheric moisture, thereby resulting a chain-to-chain crosslinking through oxygen atoms.

Uses

The phosphonitrilic halides and their derivatives have not found any major application. However, they find uses in flamer- proofing of fabrics, as plasticisers and as catalysts in the manufacture of silicones.

MAIN POINTS OF INTEREST IN PHOSPHONITRILIC COMPOUNDS OR POLYPHOSPHAZENES

There are two main points of interest in these P–N compounds:

1. The nature of the bonding is not understood. In all of these phosphazene compounds the (apparent) P–N and P–N bonds are equivalent. Their bond lengths are 1.56 – 1.59 Å, which is much shorter than the usual single bond distance of 1.77 Å.

Thus, the bonding in these compounds is not adequately represented by a system of alternate single and double bonds, nor can it be explained by $p\pi$–$p\pi$ bonding and delocalization similar to that in benzene or graphite. It has been suggested that a coordinate bond is formed between a filled sp^2 orbital on N and the empty $3d_x{}^2 - y^2$ orbital on P. This is similar to the $p\pi - d\pi$ bonding in the oxides of phosphorus except that in the phosphazenes the π bonds are delocalized over the whole molecule, giving pseudoaromatic character. There are objections to this because of the size and energy of the *d* orbitals. Alternatively the singly occupied P_z orbital on N may form a three-centre bond with the d_{xz} and d_{yz} orbitals on the two adjacent P atoms.

2. The polyphosphazenes form a very extensive series of polymers. Carbon based polymers are the most extensive, then the silicones, then the phosphazenes.

USES OF PHOSPHAZENES

There are many potential uses for the high molecular weight phosphazenes, as rigid plastics, expanded foam, and fibres since they are waterproof and fireproof, and are unaffected by petrol, oil, and solvents. They also form flexible plastics which are useful for fuel hoses and gaskets since they retain their elasticity at low temperatures. The phosphazenes are at present far too expensive for general use. Thin films of poly (aminophosphazene) are used in hospitals to cover severe burns and other extensive wounds since they prevent the loss of body fluids and keep germs out.

ANALYSIS OF PASSIVITY

The phenomenon of passivity is closely related to the protection of metals from corrosion. The term passivity was first utilised by *Schonbein* in 1836 to *describe the strange of behaviour of iron in conc, nitric acid.* Similar behaviour was reported in the case of cobalt, nickel, aluminium and chromium. Now it is commonly accepted that the phenomenon of passivity is common to all the metals.

A piece of iron when dipped in dil. HCl or H_2SO_4 evolves hydrogen gas, precipitates silver from silver nitrate and copper from copper sulphate solution. But if the same piece of iron is first dipped in conc, nitric acid or chromic acid for some time and then reacted with the above reagents, it remains is known as passivity. Such behaviour is also shown by chromium, nickel, aluminium and some other metals. Hence, "*The intertness exhibited by metal under conditions in which chemical activity is to be expected is called chemical passivity and the chemical substances which cause this phenomenon are called passivators*".

Common passivators are strong nitric acid, sulphuric acid and chromic acid. Metals of the iron group and chromium can be made passive by heating in air.

The corrossion resisting properties of stainless steel and of chromium plate are undoubtedly due to the formation of oxide films on their surfaces where exposed to air. Passivity depends upon:

(i) *The nobility of the metal.*

(ii) *The medium and the working conditions,*

(iii) *Thee initial surface state and*

(iv) *The physico-chemical properties of the covering layers, i,e, the chemical nature and structure, type of bonding, degree of cover, porosity, conductance, etc*

ALTERNATIVE DEFINITION OF PASSIVITY

Passivity or passivation may be defined as ''*phenomenon in which as metal or an alloy exhibits a much higher corrosion resistance than expected from its position in the electro-chemical series.*'' Passivity is the result of the formation of a highly protective, but very thin (about 0.0004. mm thick) and quite invisible film on the surface of metal or an alloy, which makes it more noble. This film is considered to be insoluble, non-porous and of such a ''*self-heating nature*'' that when broken, it will repair itself or re-expousre to oxidising conditions.

Examples of passive metals and alloys are *titanium, aluminium, chromium* and a wide variety of stainless steel alloys, containing chromium. All these materials exhibit outstanding corrosion resistance in certain oxidising environments,, but in reducing environment they become chemically active. Based on experiment conducted in areated-0.5 NaCl solution, the passivity of certain metals falls in the following order.

$$Ti \longrightarrow Al \longrightarrow Cr \longrightarrow Be \longrightarrow Mo \longrightarrow Mg \longrightarrow$$

$$Ni \longrightarrow Co \longrightarrow Fe \longrightarrow Mn \longrightarrow Zn \longrightarrow Cd \longrightarrow$$

$$Sn \longrightarrow Pb \longrightarrow Cu.$$

It may be pointed here that *passivation is not a constant state, but exists only uncertain environmental condition, which tends to maintain thin protective oxide film on their surface.* In the presence of oxygen, the oxide film will be automatically repaired whenever any damage occurs, but in its absence the passive metals and alloys become chemically active and are rapidly corroded. For example, the austentic stainless steels are quite good resistant to the action of aerated dilute sulphuric acid, but show low resistance in air-free acid.

THEORIES OF PASSIVITY

Many theories have been advanced to explain the nature of passivity but none of them can yet give a complete and satisfying explanation of all the manifestations of this phenomenon. However, all theories fall into the four groups:

1. Gaseous Film Theory

According to the theory, a thin film of the gas is said to adhere to the said surface of the metal and thus rendering it passive. The gaseous film may be of oxygen or of nitric oxide or may be of any gas. A close connection has been found to exist between chemical passivity and photo-electric behaviour of metals which supports this theory. However, the main-drawbacks of this theory are:

(1) It fails to explain the passivity of lead in the presence of sulphuric acid.

(2) It also fails to explain the fact why passive iron is rendered active by scratching the surface.

2. Physical Theories

(a) Catalytic Theory : According to this theory, pure iron is normally passive but may be made active by the presence of a catalyst like hydrogen or hydroxyl ions which increases the solution pressure of the metal. The experiments performed by Lambert and Thomson support this theory who showed that pure samples of iron were passive.

Again, this theory is not able to explain passivity fully.

(b) Valence Theory : According to this theory, active iron is in the bivalent state while passive iron is in the trivalent state.

This theory was proposed by *Kruger* and *Finkelstein* and again failed to explain all types of passitivites met within practice.

(c) Allotropic Theory : According to this theory active metal is converted into a passive metal due to a stable allotropic modification. This theory does not explain any phenomenon because it simply substitutes another name for passivity.

3. Oxide Layer Theory : The oxide layer theory owes its origin to the first hypothesis of Faraday (1936). According to this theory:

(i) *The metal surface was oxidised or that atoms of the metal were in such a relation to the oxygen as to be equivalent to an oxidation.*

(ii) *The surface oxide film would be so thin as to be invisible.*

(iii) *The oxide layer could not be dissolved by acids.*

As a result of metal being completely shielded by the layers of oxide, it could no longer be dissolved anodically, be attacked by acids or could replace the metals from solutions. 'In other words, the metal becomes passive due to the formation of an oxide film on its surface.

Objections : This theory held its ground for many years until *Hittorf* (1898) showed that:

(i) *The properties of oxide responsible for passivity must be different from those of any known* oxide of the metal.

(ii) *As the light reflecting powers of a passive anode and a cathode of the same metal* (*which is of course in the active condition*) *are the same, it suggested that there could be no oxide film on the former, i.e., anode.*

Most of the above mentioned difficulties have been overcome in recent years. It was experimentally proved that an oxide film was formed and in one case it was actually isolated.

It is now observed that in certain cases films other than that of the oxides are also the causes of passivity. Phosphate, silicate and nitride films have been detected. Thus, a slight modification of the oxide layer theory is required which may be summarised as below:

"*Any protective film formed on the surface during the chemical action is sufficient to cause passivity.*"

Thus, the oxide film theory must modified to be termed as protective layer theory. It is important to remark here that oxide film is formed in most of the passivity cases.

Evidences. The various evidences in favour of this theory are:

(i) *In case, the thin film of protective layer is removed mechanically or chemically, the metal is rendered active again.*

(ii) *When passive iron is heated with cone nitric acid up to* 75 °C, *it becomes active again and begins to exhibit its usual reactions. It is due to the fact that ferric oxide on the surface of passive iron begins to dissolve in cone.* HNO_3 *at* 75°C.

(iii) *It has been shown that polarised light* (*and not ordinary light*) *is not reflected to the same extent by active and passive forms of a given metal. It shows that some layer is formed on the metal surface.*

(iv) **Evan** *succeeded in dissolving away the inner portion of passive iron leaving an almost colourless film of oxide. This film is quite invisible on the passive iron but can be seen when separated from it.*

(v) *Iron is rendered passive not only by nitric acid but also by other oxidising agents like fused nitrates, chromic acid, etc. confirming that passivity is due to the formation of oxide film on the surface of the metal.*

(vi) *A large number of other metals like aluminium, chromium, bismuth, cobalt and nickel become passive with cone, nitric acid due to the formation of protective oxide films on their surfaces.*

Thickness of the Film : The rate of formation of oxide film in passivity has been investigated by a number of workers. They proved that the rate of film growth is inversely proportional to the thickness of the film and obeying the following equation:

$$\frac{dy}{dt} = ky^{-1} \qquad ...(1)$$

where y is the film thickness, t is the time and k is a constant.

On integrating Eq. (1) we get

$$y^2 = 2kt + \text{constant} \qquad ...(2)$$

Equation (2) is the equation of a parabola. Thus, the rate of film growth must obey the parabolic law. Experimentally, it has been found in many cases but in some cases there are a number of complications. Thus, a film is not necessarily uniform throughout its thickness. Generally, the thickness is of the order of zero to 10^{-8} Å.

4. Theory of Electronic Configuration : Various attempts to reconcile the numerous theories with 'the protective layer led to the electronic configuration theory.

On the suggestion of *Russel. Siwinne* and *Sbursi* the theory of electronic configuration was developed quantitatively by *H.H. Uhlig.* The main postulates of this theory are:

(i) *Passive metals belong to the transition group of the periodic table. Examples of this are: Cr, Ni, Co, Fe, Mo and W.*

(ii) *Atoms of these transition metals have an incomplete filling of the d-orbitals in their crystalline structure. These incomplete d–orbitals tend to fill up with electrons.*

(iii) *The passive conditions correspond to that with incomplete d-orbitals and the active condition to that when d-orbitals are filled up.*

Thus, the adsorption of oxygen or oxidising substances to the metallic surface causes passivity since they behave as electron adsorbers with no tendency to lose electrons to the surface atoms of the metal.

On the other hand, when dissolved hydrogen and other catalytic substances are adsorbed on the metal surface, they have the tendency to furnish electrons to metal atoms and fill up d-orbitals. Hence, this favours the stabilisation of the active state.

IS PASSIVITY AN UNIVERSAL PHENOMENON

Recent development in the phenomenon of passivity indicates that it is not limited to only metals like iron, chromium, aluminium etc. but almost to all metals.

APPLICATIONS OF PASSIVITY

The phenomenon of passivity though undesirable has proved to be advantageous in a number of cases. Protective oxide films on the metals are very useful in preventing corrosion. For example:

(i) *Cast iron retorts are used for the manufacture of nitric acid by heating potassium nitrate with cone, sulphuric acid. The oxide film formed on the iron surface of the retort due to passivity protects iron from corrosion.*

(ii) *Aluminium is generally used in air-craft industry The oxide film is deposited on aluminium surface to resist corrosion.*

The aluminium article is made as an anode and dipped in a cell containing 3% chromic acid solution. When electricity is passed, and excellent protective oxide film is deposited which resists corrosion up to a remarkable extent.

(iii) *Similarly, stainless steel is covered by an oxide film to prevent its corrosion. This type of steel is used in the manufacture of automobile parts.*

(iv) *Metals which are susceptible to corrosion are made passive by alloying with one or more metals which are passive. For example, iron is rendered passive by alloying with any of the transition metals such as chromium, nickel and molybdenum.*

ELECTRO-CHEMICAL PASSIVITY

Electro-chemical passivity is now generally recognised as the same phenomenon as the chemical passivity. The only difference is that electro-chemical passivity is brought about by anodic oxidation. However, it may be defined as follows:

In electro plating or electrolysis, the anodes sometimes become passive if a high current density is passed or when a strong oxidising agent is present in solution. This is termed as electro-chemical passivity. As this type of passivity is important in electro-chemical technical processes, it has been studied by a large number of scientists. Some important points about electro-chemical passivity are.

(i) *Electro-chemical passivity is shown by all metals under suitable conditions but depends also on electrolytic solution.*

(ii) *Electro-chemical passivity may be caused by a small current passing for a long time or heavy current for a short time.*

(iii) *The electrical energy is utilised in performing chemical reactions and also used for causing passivity. Thus, the electric energy is wasted.*

Electro-chemical passivity is a technically important evil in electro-chemical industries such as electro-refining and electro- plating because a large portion of electric energy is wasted when anode becomes passive. Inspite of this misuse, an important use of that it is being used for deposition of an oxide film on aluminium anode which is dipped into a cell containing 3% chromic acid. This layer protects aluminium form corrosion and is used in aircraft industry.

Mechnaical Passivity

In certain cases, the metal becomes passive due to the formation of visible, insoluble and comparatively thick oxide film. This type of passivity is called mechanical passivity. For example: lead dioxide on a lead surface in dil. sulphuric acid. Mechanical passivity is not fundamentally different from the forms of passivity already discussed. Some important points about mechanical passivity are:

(ii) *The film deposited on the anode increases the resistance of a cell.*

(iii) *Mechanical passivity has been observed with iron, cobalt, nickel and other metals in alkali hydroxide solution along with thallium, antiomony and bismuth in aqueous sulphuric acid.*

Mechanism : The possible mechanism of film formation is:

(i) *It can be readily formed when the solubility product of certain poorly soluble substance (tike lead sulphate) is exceeded.*

(ii) *Secondly, it may be produced by electrophoresis of colloidal particles present or formed in the bulk of the solution.*

Use: Mechanical passivity is utilised for the manufacture of a low power electrolytic rectifier from aluminium since it has a peculiarity of offering a high resistance to current in one direction and a much lower one in the other.

ANALYSIS OF CORROSION

During this process mostly oxides are formed and depending upon the presence of impurities sulphides, carbonates, sulphates, etc. may also be formed.

The process by which the metals have tendency to go back to their combined state is called corrosion. As soon as the metals are extracted from their ores, the reverse process begins, *i.e.,* the nature tries to convert them back into the form in which they occur.

Modern Definition : Recently corrosion has been defined *as "the deterioration of a substance because of its reaction with its environment.*

Under this definition, the term corrosion may by applied to gradual transformation of the metals into compound state, to weathering of timbers and concrete, to the leading of glass, to the cracking of plastics, etc. However, metals are still considered to be the major materials subject to corrosion and the treatment here will be confined to processes involving metals.

ECONOMIC ASPECTS OF CORROSION

The economic aspects of corrosion are two fold:

(i) *First is the cost of replacement of materials and equipment lost through corrosion. The current estimate is nine billion dollars per year for the replacement of pipes, building products, autoparts. water heater and other items subject to corrosion.*

(ii) *The second aspect is the limitation of progress in certain areas because of corrosion problems. Better chemical processing and refining operations, production of oil from certain strata and conquest of oceans are all areas where progress hings to some extent, on the solution of corrosion problems.*

Most corrosion costs are due to corrosion in eletrolytes or wet corrosion.

TYPES OF CORROSION

Corrosion of the metals occurs by the attack of surrounding environment/medium on the surface of metal. This attack occurs in two ways, viz., (i) *Direct chemical corrosion or dry corrosion; and* (ii) *Electro-chemical corrosion or wet corrosion or immersed corrosion.*

CHEMICAL CORROSION

The type of corrosion occurs mainly through the *direct chemical action of environment/atmospheric gases such as oxygen, halogens, hydrogen sulphide, sulphur dioxide, nitrogen or anhydrous inorganic liquid with metal surfaces in immediate proximity.* The extent of corrosion depends upon the :

(i) Chemical affinity between the corrosive environment and solid metals, and

(ii) Ability of reaction product on metal surface to form a protective film.

There are *three* main types of chemical corrosion :

1. *Oxidation corrosion* is brought about by the direct action of oxygen at low or high temperatures on metals, usually in the absence of moisture.

At ordinary temperatures metals, in general, are very slightly attacked. However, alkali metals (Li, Na, K, Rb, etc.) and alkaline earths (Be, Mg, Ca, Sr, etc.) are even rapidly oxidised at low temperatures. At high temperatures, almost all metals (except Ag, Au, Pt) are oxidised.

The reactions in the oxidation corrosion are:

$$2M \longrightarrow \underset{\text{Metal ions}}{2M^{n+}} + 2ne^- \quad \text{(Loss of electrons)}$$

$$\frac{n}{2}O_2 + 2ne^- \longrightarrow \underset{\text{Oxide ions}}{nO_2} \quad \text{(Gain of electrons)}$$

or

$$2M + \frac{n}{2} \longrightarrow \underset{\text{(Metal oxide)}}{2M^{n+} + nO^{2-}}$$

Mechanism : Oxidation occurs first at the surface of the metal and the resulting metal oxide scale forms a barrier that tends to restrict further oxidation. For oxidation to continue, either the metal must diffuse outwards through the scale to the surface or the oxygen must diffuse through the scale to the underlying metal. Both transfers occur but the outward diffusion of metal is, generally much more rapid than the inward diffusion of oxygen, since the metal ion is appreciably smaller than the oxygen ion and, consequently, is of much higher mobility.

The *nature of the oxide formed* plays an important part in oxidation corrosion process.

$$\text{Metal + Oxygen} \longrightarrow \underset{\text{(Corrosion product)}}{\text{Metal Oxide}}$$

When oxidation starts, a thin layer of oxide is formed on the metal surface and the nature of this film decides the further. If the film is:

(i) **Stable :** *A stable layer is fine-grain in its structure and it can get adhered lightly to the parent metal surface.* Hence, such a layer can be of impervious nature (*i.e.,* which cuts off penetration of attacking oxygen to the underlying metal).

In other words, it amounts to sealing the metal surface from any further contact with atmospheric oxygen. Such a film behaves as protective coating in nature, thereby shielding the metal surface The oxide films on the surface of metals aluminium, tin, lead, copper, etc, are stable, adherent and impervious in nature. Consequently, further oxidation corrosion is prevented.

(ii) **Unstable,** *i.e.,* the oxide layer formed decomposes back into the metal and oxygen.

$$\text{Metal oxide} \rightleftarrows \text{Metal + Oxygen}$$

Consequently, oxidation corrosion is not possible in such a case. Thus, silver, gold, platinum do not under go oxidation corrosion.

(iii) **Volatile**, *i.e., the oxide layer volatilizes as soon as it is formed, thereby leaving the underlying metal surface exposed for further atack.* This causes rapid and continuous corrosion, leading to excessive, corrosion *e.g.,* molybdenum oxide (MoO_3) is volatile.

(iv) **Porous**, *i.e., having pores or cracks.* In such a case, the atmospheric oxygen have access to the underlying surface of metal, through the pores or cracks of the layer, thereby the corrosion continues unobstructed till the entire metal is completely converted into its oxide.

Pilling-Bedworth rule on protective and non-protective nature of oxide layer : *According to it ' 'an oxide is protective or non- porous, if the volume of the oxide is at-least as great as the volume of the metal from which it is formed."* On the other hand, "*if the volume of the oxide is less than the volume, the oxide layer is porous (or non-continuous) and hence is non- protective, because it cannot prevent the access of oxygen to the fresh metal surface below.*

"Thus, alkali metals and alkaline earth (like Li, K, Na, Mg,) form oxides of volume less than the volume of metal from which oxide is formed. In other words, volume of oxide formed is insufficient to cover completely the compartively larger surface area of the metal.

Consequently, the oxide layer faces stress and strains, thereby developing cracks and pores in the structure. Porous oxide scale permits free access of oxygen to the underlying metal surface (through cracks and pores) for fresh action and *thus corrosion continues non-stop.*

On the other hand, metals *like aluminium forms oxide* (on oxidation), *whose volume is greater than the volume of metal* (Al). *Consequently, an extremely tightly-adhering non-porous layer is formed.* Due to the absence of any pores or cracks in the oxide film, it forms a barrier for further action and the rate of oxidation is rapidly decreased to zero.

2. *Corrosion by other gases :* Among the other dry gases which cause corrosive effect on the metals are SO_2, CO_2 ,Cl_2, H_2S, F_2, etc. The extent of corrosive effect depends mainly on the chemical affinity between the metal and the gas involved. Just as in the case of oxidation corrosion by air, the degree of attack in such eases also depends on the formation of protective or non-protective films on the metal surface.

(a) *If the film formed is* protective or non-porous (*e.g., AgCl film resulting from the attack of Ch on Ag*), *the intensity or extent of attack decreases, because the film formed protects the metal from further attack.*

(b) *If the film formed is* ***non-protective or porous****, the surface of the whole metal is gradually destroyed. For example, dry* Cl_2 *gas attacks on tin* (S_n) *forming volatile* $SnCl_4$ *which vola-talizes immediately from the tin surface, thereby leaving fresh metal surface for further attack. Similarly, in petroleum industry,* H_2S *at high temperature attacks steel forming a* FeS *scale, which interferes with normal operations.*

Corrosion by Hydrogen

(a) *Gases like hydrogen also attack metals at ordinary temperature, It is known as '****hydrogen embrittlement****' Hydrogen embrittlement is due to the formation of atomic hydrogen as the result of chemical or electrolytic action occurring at a metal surface under specific environment. For example, the presence of aqueous solution of* H_2S *in the system, causes evolution of atomic hydrogen at the iron surface.*

$$Fe + H_2S \longrightarrow FeS + 2H \text{ (atomic hydrogen)}$$

This evolved atomic hydrogen diffuses readily into thc metal and collects there, in voids, etc., where it recombines to form molecular hydrogen.

$$H + H \longrightarrow H_2$$

The process of formation of atomic hydrogen, diffusion into the metal, collection into metal voids and finally formation of molecular hydrogen continues and the pressure of hydrogen inside the metal voids continues to increase, till this pressure becomes too high to cause blisters and fissures.

This causes, consequently lowering of strength and ductility' of the metal. In other words, the metal becomes weak.

(b) *At high temperatures, hydrogen attack is also due to diffusion of atomic hydrogen into metal. The atomic hydrogen is formed by the thermal dissociation.*

$$H_2 \xrightarrow{\text{Heat}} H + H \quad \text{(Thermal dissociation)}$$

The atomic hydrogen thus formed at high temperature is highly chemically active and readily combines with, C, S, O, or N usually present in small amounts in metals. For example, atomic hydrogen readily combines with C in steel at high temperatures forming high pressure methane (CH_4) gas, which causes, cracking, blistering etc. in the metal. As a result of this, the metal becomes brittle and weak. This phenomenon is known as 'decarburization'.

3. *Liquid metal corrosion is due to chemical action of flowing liquid metal at high temperatures on solid metal or alloy.* Such corrosion has been found to occur in devices used for nuclear power. The corrosion reaction involves either (i) dissolution of a solid metal by a liquid metal or (ii) internal penetration of the liquid metal into the solid metal. Both these modes of corrosion cause weakening of the solid metal.

WET OR ELECTRO-CHEMICAL CORROSION OR IMMERSED CORROSION

This type of corrosion occurs: (i) where a conducting liquid is in contact with metal or (ii) when two dissimilar metals or alloys are either immersed or dipped partially in a solution.

The *corrosion occurs due to the existence of separate 'anodic' and 'cathodic' areas parts between which current flows through the conducting solution. At anodic area oxidation reaction* (*i.e., liberation of free electrons*) *takes place.* Consequently, there is a tendency at anode to destroy the metal by either dissolving it as ion or to revert to a combined state such as oxide, etc. *Hence, corrosion always occurs at anodic areas.*

At anode:

$$\underset{\text{Metal}}{M} \longrightarrow \underset{\text{Metal ion}}{M^{n+}} + ne^-$$

$$\underset{\text{Metal ion}}{M^{n+}} \longrightarrow \text{Dissolves in solution.}$$

or

$$\longrightarrow \text{Forms compound such as oxide.}$$

On the other hand, *at cathodic areas, reduction reaction* (*i.e., gain of electrons*) *takes place.* Usually, cathode reactions do not affect the cathode part of the metal, since most metals cannot be further reduced. So usually at cathodic part, dissolved constituents in the conducting medium accept the electrons to form some ions (OH^-, O^{2-}). The metallic ions at anodic part and some other ions formed at cathodic part diffuse towards each other through conducting medium and form a corrosion product somewhere between anode and cathode. The electrons set free at the anode flow through the metal and are finally consumed in the cathodic reaction.

Thus, we may sum up that elctro-chemical corrosion involves :

(i) *the formation of anodic and cathodic areas or parts in contact with each other;*

(*ii*) *presence of a conducting media.*

The rate of electro-chemical corrosion depends upon the nature of corrosion product:

(i) *If the corrosion product goes into solution or is evolved as a gas, the corrosion reactions go on unaltered or unabated.*

(ii) *If the corrosion product is an insoluble compound, then it is likely to cover the metal surface, thereby reducing the rate of subsequent corrosion. However, if the insoluble compound is formed at a distance from the corroding anodic area, it will have little protective value.*

CORROSION BY GASEOUS ENVIRONMENTS

In this type of corrosion, the metal is surrounded by gases present in the surrounding environment. This type produces two effects on the metal:

(i) *Metal is consumed, and*

(ii) *The metal properties are changed.*

The factors determining the rate of corrosion are different for different conditions:

(A) *Metal sheltered from rain or*

(B) *Metal exposed to outdoor atmosphere and it is exposed to rain periodically.*

(A) Metals Sheltered from Rain : There are large number of factors which have the maximum influence.

(a) Moisture : The amount of moisture present in the atmosphere is very important factor. This is recognised in the "*Principle of critical humidity*" which applies to most of the metals. *The percentage humidity below which the corrosion is negligible and above which it is appreciable is called the "critical humidity value"*. For iron, 50 – 60% is called the critical humidity value of iron. For iron there is one more value of critical humidity viz., 80 – 90%. At this value, corrosion becomes very serious and the surface of the metal gets covered with red rust.

(b) Suspended impurities in air : Suspended impurities have a great effect on the rate of corrosion of metals. For example, in case of iron the formation of red rust can be prevented if it is allowed to come in contact with air which is free from any suspended impurities. From the point of view of corrosion, the suspended particles in air may be classified under three broad headings:

(i) ***Intrinsically active particles :*** *These are active particles which can bring about the corrosion by themselves. For example sodium chloride near sea and ammonium sulphate in town air.*

(ii) ***Intrinsically neutral particles :*** *These are themselves chemically inactive but have a good capacity for adsorbing gases from the atmosphere. For example the amorphous carbon present in the atmosphere can adsorb many gases and can, therefore, enhance the rate of corrosion.*

(iii) ***Neutral particles :*** *These are the particles which are neither chemically active nor capable of adsorbing active gases of the atmosphere. Thus, such particles have no influence on the rate of corrosion, i.e., crystalline silica.*

(c) Effect of active gases: The rate of corrosion is seriously influenced by the presence of active gases in atmosphere. Examples of active gases are CO_2, H_2S, SO_2, etc.

Traces of sulphur dioxide present in the atmoshphere yield basic copper sulphate film on the surface of copper.

Carbon dioxide plays a fundamental part in the corrosion of metals like iron to an unlimited extent.

(d) Nature of the primary film of corrosion : If the film is continuous and adheres to the surface, it stops the further corrosion and the film is said to be of the *protective type.* If the film is not continuous or it may leave the surface, then the corrosion is occurring rapidly. For example, heavy metals and aluminium form oxide films on their surfaces. The nature of these films is continuous and thus protects the metal from further corrosion. Alkali and alkaline earth metals form porous layers and thus do not offer any resistance to further corrosion.

(B) Metals Exposed to Outdoor Atmosphere: This type of corrosion takes place in the same manner as in indoor atmosphere. The only additional factor which comes into play is rain. Previously it was thought that the effect of the rain on corrosion was generally adverse but now it is known that rain is not necessarily a disadvantage because it may help to develop a product having protective qualities as in the case of copper. Thus, it helps to retard the corrosion. However, rain may have the following effects on corrosion:

(i) *It may remove a corrosion product as in the case of nickel.*

(ii) *Rain may help to develop a product having protective qualities as in the case of copper.*

(iii) *The rate of attack may be increased by the breakdown or removal of an otherwise protecting coating.*

Theory of Corrosion by Gaseous Environments

Atmospheric corrosion may be explained by the theory which involves the following postulates.

(a) *The gas is dissolved in the metal as molecules, atoms or ions to form solution. This is followed by diffusion of the gas in metal.*

(b) *A solid or liquid reaction product is formed on the surface. Reaction continues by the diffusion of the gas or metal atoms or ions through the surface film.*

In the case of oxygen, oxide film is obtained; in the case of hydrogen sulphide, sulphide is formed; in the case of CO_2 carbonate is formed.

(c) *The film, if it is non-porous, affords protection to the metal from further attack. The metal is then said to be in the passive state.*

(d) *When the film is porous, it causes serious attack and results in serious corrosion.*

Three important types of corrosion products are formed on the surface.

(a) *Products formed on the surface may be unstable.*

$$M + 1/2O_2 \longrightarrow MO$$

$$MO \longrightarrow M + 1/2O_2$$

In such cases, there will be no corrosion as in the case of noble metals, i.e., platinum oxide formed is unstable.

(b) *Products formed may be volatilised at high temperatures. In such cases excess of corrosion takes place. This type of corrosion occurs in case of molybdenum, tungsten, etc.*

(c) *Product formed may be stable. Most of the metals fall into this category. This may further be divided into three types:*

(i) *The product film, which cannot completely cover the metallic surface and forms a porous film, enables the gases to diffuse and attack further. This causes serious corrosion.*

(ii) *The product film may be non-porous and built up to certain thickness. In such cases rate of corrosion decreases with time. This type of corrosion occurs in most of the metals.*

(iii) *In some cases, non–porous product film may be brittle and may crack after some time. In such cases, therefore, deeper corrosion may occur,*

$$2Al + 3/2\ O_2 \longrightarrow Al_2O_3$$

$$2Fe + 3/2\ O_2 \longrightarrow Fe_2O_3.$$

IMMERSED CORROSION

There are four types:

(I) Chemical corrosion.

(II) Underwater corrosion.

(III) Underground corrosion.

(IV) Concentration cell corrosion.

Let us discuss these one by one.

1. Chemical Corrosion

This type of corrosion is the result of electro-chemical-CHEMICAL or chemical reaction between a metal and its liquid surrounding. There are large number of factors which affect the rate of corrosion of metals under such conditions:

(a) Concentration of salts : Increase in the concentration of salts increases corrosion. However, beyond a certain concentration the rate of corrosion is decreased. This decrease in the rate of corrosion is thought to be due to lower solubility of oxygen in the more concentrated solution.

(b) Depth of immersion : The rate of corrosion is directly proportional to the depth of immersion. The more the metal immersed the more the corrosion.

(c) Concentration of dissolved oxygen : The concentration of dissolved oxygen is directly proportional to the rate of corrosion. Increase in the concentration of dissolved oxygen increases the rate of corrosion.

(d) Temperature. Corrosion increases with increase in temperature.

(e) Nature and concentration of cation of the medium : In most of the cases, corrosion in a liquid medium is acid corrosion, *i.e.,* it is type of corrosion due to substitution of hydrogen by a metal in the corroding medium. When metallic salts are present in the corroding medium then the metal from the corroding medium is substituted by the corroding metal.

The concentration of cations, *i.e.,* metal ions present in corroding medium, influences the acid corrosion. For example, concentrated solutions of alkali carbonates do not corrode iron but dilute ones corrode it to some extent.

(f) Nature and concentration of anions of the medium : All the anions do not influence the corrosion. However, traces of halides present in the corroding medium affect the corrosion as these help in dissolving the protective film formed on the metal to be corroded and thus increases corrosion.

(g) Nature of corrosion products : On the basis of their physical nature and behaviour, the corrosion products may fall into three categories:

(i) *Those which are soluble.*

(ii) *Those which are fixed to the metal and are insoluble.*

(iii) *Those which do not fix to the metal and are discontinuous.*

In the first case, corrosion will continue until all the metal is consumed; in the second case, the metal will be protected from further corrosion: in the third case the corrosion will be serious under certain conditions.

(h) Electrical conductivity of the corroding medium : Electrical conductance of a corroding medium affects the corrosion. If the corroding medium has a low conductance, the rate of corrosion is slow whereas if the conductance is high, the rate of corrosion will be fast.

(i) ***Position of the metal in the E.M.F. series :*** *The more basic the metal, the more readily it will corrode.*

(j) pH : The lower the pH, the greater the corrosion.

(k) Presence of colloids : J. Newton Friend reported that the presence of colloids inhibits corrosion.

(I) Presence of bacteria : In some cases bacteria influence the corrosion of metal.

2. Under-water Corrosion

This type of corrosion includes corrosion in water pipes of all types, heating systems, steam boilers, etc. The water is in sufficient amount whereas the air, in such cases, is limited.

3. Under Ground Corrosion

Corrosion of pipes and other underground structures and the corrosion of ferrous metals in concrete come under this category. It is sometimes complicated due to the presence of various salts in the soil. Conditions existing in the soil may vary from those similar to atmospheric exposure to those comparable to complete immersion. The corrosiveness of a soil depends upon : (i) *its acidity*, (ii) *degree of aeration,* (iii) *electrical conductivity*, (iv) *its moisture and salts content*, (v) *presence of bacteria and micro- organism, and* (vi) *soil texture.*

Soil corrosion is purely electro-chemical in character. *The texture of a soil is determined by the percentages of particles of various sizes.* According to particle size, soils are classified as : (i) *gravels and stones*, (ii) *fine gravel,* (iii) *sand*, (iv) *silt, and* (v) *clay.*

(a) Gravelly or sandy soils are very porous and are, therefore, usually strongly-aerated : If a metal pipe is hurried in such a soil, corrosive conditions are similar to those under wet condition, and the corrosion rate in such a case will, evidently, be governed by amount of moisture content in the soil.

(b) In water-logged soils, the amount of free oxygen available is very small, but various bacteria and micro-organism can grow, which may *lead to micro-biological corrosion.* In acidic soils, the mechanism of soil corrosion will be similar to the hydrogen evolution type. The rate of such corrosion depends on : (i) the pH (or acidity) of the soil, (ii) the presence of salt, and (iii) the presence of oxygen, etc. The presence of oxygen facilitates the evolution of hydrogen and, hence, accelerates the rate of attack.

(c) Intermediate character soils (fine gravel, sand and silt containing) may produce a localized and intense corrosion. If air pockets are present (either originally in the soil or are produced artificially after laying of pipeline), differential aeration corrosion may occur in different parts of pipeline.

4. Concentration Cell Corrosion

This type of corrosion is due to electro-chemical attack on the metal surface, exposed to an electrolyte of varying concentrations or of varying aeration. This may be the result of local differences in metal-ion concentrations, caused by local temperature differences or inadequate agitation or slow diffusion of metal ions produced by corrosion.

Differential aeration corrosion is the most common type of concentration cell corrosion and it occurs when one part of metal is exposed to a different air concentration from the other part. This causes a difference in potential between differently aerated areas. It has been found experimentally that "*poor-oxygenated parts are anodic*". Consequently, a differential aeration of metal causes a flow of current, called the *differential current.*

Differential aeration accounts for the *corrosion of metals, partially immersed in a solution, just below the water-line.* Thus, if a piece of a metal (say Zn) is partially immersed in a dilute solution of a neutral salt (say NaCl) and the solution is not agitated properly, then, *the parts above and closely adjacent to the water-line are more strongly aerated, because of the easy access of oxygen and hence become cathodic. On the other hand, parts immersed to greater depth, which have less access of oxygen, show a smaller oxygen concentration and thus become anodic.* So, a difference of potential is created, which causes a flow of current between the two differentially aerated areas of the same metal. Zinc will dissolve at the anodic areas,

$$Zn \longrightarrow Zn^{2+} + 2e^- \qquad \text{(oxidation)}$$

and oxygen will take up electrons at the cathodic areas to form hydroxyl ions.

$$1/2\ O_2 + H_2O + 2e^- \longrightarrow 2OH^- \qquad \text{(reduction)}$$

The circuit is closed by migration of ions through the electrolyte and a flow of electrons takes place, through the metal from anode to cathode.

In a similar way, iron corrodes under drops of water (or salt solution). Areas covered by droplets, having less access of oxygen, become anodic in respect to the other areas freely exposed to air.

From the above, it is clear that *oxygen concentration cell increases corrosion, but it occurs where the oxygen concentration is lower.* This generalization is significant.

(i) *Corrosion may be accelerated in apparently inaccessible places, because the oxygen-deficient areas serve as anodes and, therefore, cracks or crevices serve as foci of corrosion.*

(ii) *Similarly, corrosion is also accelerated under accumulation of dirt, sand, scale or other contamination. This is because accumulation of rust or scale or sand, etc. restricts the access of oxygen and establishes an anode to promote still greater accumulation. The result is localized corrosion, due to non-uniform corrosion.*

(iii) *Likewise, metals exposed to aqueous media corrode under blocks of wood or pieces of glass, which screen that portion of metal from oxygen access. The differential areation type of corrosion is a localized attack on some oxygen-deficient areas and results in characteristics localized pitting. This attack becomes more intensified with the time, because the corrosion products accumulate around a small anodic area thus making inaccessibility of that part more effective.*

THEORIES OF IMMERSED CORROSION

Various theories have been put forward to explain the phenomenon of immersed corrosion but the accepted theory is the modern electrochemical theory.

Most of the theories are concerned with iron metal as it is the most common metal to undergo immersed corrosion.

(a) Carbonate Formation Theory or Acid Theory : According to this theory, a little amount of an acid is necessary for corrosion to take place. Even carbon dioxide dissolved in water is enough. Initial corrosion is started by the joint action of carbon dioxide, water and oxygen.

$$CO_2 + H_2O \longrightarrow H_2CO_3$$
$$Fe + H_2CO_3 \longrightarrow FeCO_3 + H_2$$
$$H_2 + \frac{1}{2}O_2 \longrightarrow H_2O$$

$$4FeCO_3 + 10H_2O + O_2 \longrightarrow 4Fe(OH)_3 + 4H_2CO_3$$

In the last reaction, carbonic acid H_2CO_3 is regenerated which may further attack the metal and thus corrosion proceeds. Excess of oxygen helps corrosion. When atmosphere contains SO_2, H_2S, etc, they may react with iron.

Limitations: Various limitations are: •

(i) *This theory is not supported by practical observations.*

(ii) *This theory does not explain why corrosion proceeds in the complete absence of carbon dioxide and even in alkaline solution.*

(b) Peroxide Theory : This theory is due to Golding and is known as Golding's chemical theory. According to this theory, iron reacts with water to liberate hydrogen which will be changed to hydrogen peroxide by dissolved oxygen.

$$Fe + 2H_2O \longrightarrow Fe(OH)_2 + H_2$$

$$H_2 + O_2 \longrightarrow H_2O_2$$

The hydrogen peroxide oxidises $Fe(OH)_2$ to $Fe(HO)_3$.

$$2Fe(OH)_2 + H_2O_2 \longrightarrow 2Fe(OH)_3$$

or it may react with metallic iron and changes it to $Fe(OH)_3$.

$$Fe + H_2O_2 \longrightarrow Fe(OH)_2$$

Limitations

Various limitations are:

(i) *This, theory fails completely because pure hydrogen peroxide has no perceptible action on iron.*

(ii) *It was also reported that the reagents which destroy hydrogen peroxide do not prevent corrosion.*

(iii) *This theory is not much supported by practical observations.*

(c) Oxygen Theory

$$4Al + 3O_2 \longrightarrow 2Al_2O_3$$

$$2Sn + O_2 \longrightarrow 2SnO$$

These oxides have the same colour as that of the metals. But when copper metal is attacked, it gets changed to copper oxide which is black. We say that copper is tarnished. However, this type of reaction cannot proceed in the absence of moisture.

Limitations

(i) *This theory is not supported by practical observations.*

(ii) *This theory cannot explain corrosion in all cases.*

(b) Electro-chemical Theory : This theory explains the corrosion fully. According to this theory the chemically non-uniform surfaces of metals behave like small electric cells in presence of water containing dissolved oxygen and carbon dioxide. Thus corrosion occurs as a result of local action. The main points are:

(i) *The covered end of the metal surface acts as the anode where iron forms* Fe^{2+} *ions by losing its electrons.*

$$Fe \longrightarrow Fe^{2+} + 2e^-$$

(ii) *A less eletropositive metal like copper which is present as an impurity conducts away the electrons and develops the cathode where hydroxyl ions are formed.*

$$2H_2O + 2e^- \longrightarrow H_2 + 2OH^-$$

The hydrogen so liberated may get accumulated at the cathode and cause polarisation which might bring a stop to corrosion, but the atmospheric oxygen acts as a depolariser.

$$H_2 + 1/2\ O_2 \longrightarrow H_2O$$

(iii) *Now, there exists two oppositely charged ions, i,e., cathodic and anodic, on the surface of the metal and a galvanic flow of ions may take place and then the two oppositely charged ions may combine to form a compound.*

$$Fe^{2+} + 2OH^- \longrightarrow Fe(OH)_2$$

Ferrous hydroxide is then converted to basic carbonate by the carbon dioxide and oxygen present in the atmosphere.

$$4Fe(OH)_2 + 2H_2O + O_2 \longrightarrow 4Fe(OH)_3$$

$$Fe(OH)_2 + CO_2 \longrightarrow Fe(OH)(HCO_3)$$

$$Fe(OH)_3 + CO_2 \longrightarrow Fe(OH)_2(HCO_3)$$

From the above theory, it follows that

"*Any corrosion process can be related to an elctro-chemical process in which cathode and anode are formed on the metal surface and an electrolyte (water or acid or salt solution) must be present to permit ionic flow to form the corrosion products.*"

When the elements are arranged in the e.m.f series of order of decreasing activity, it is observed that higher the element in the series, the more easily it gets corroded. Actually it is always the case. An interesting example is aluminium. Although it is placed near top of the series yet it resists atmospheric corrosion effectively. This is due to the formation of an oxide film on its surface which prevents further corrosion.

(e) Differential Aeration Theory : Recently, **Evans** has established that a major cause of rusting is due to differential aeration. A cell is set up between iron dipping in oxygen-free water and iron in oxygen containing water and current flows, the non-aerated part of iron acting as anode and the aerated one as cathode. Ferrous ions produced at the anode diffuse in water and combine with hydroxyl ions produced at the cathode forming ferrous hydroxide which due to aeration gets converted into the ferric hydroxide. This theory is being discussed these days and it well take some time in establishing it; validity.

MICROBIOLOGICAL CORROSION

Corrosion caused by the metabolic activity of various micro-organisms, is called **"microbiological corrosion"**. The micro-organisms can develop in an environment with or without oxygen and are classed as *aerobic or anaerobic*. Important corrosion failures due to activity of micro-organisms are:

(1) *Sulphate reducing bacteria,* like '*Sporovibrio desulphuricous*, are responsible for anaerobic corrosion of iron and steel, because these can grow only under anaerobic conditions. This requires in addition to oxygen, adequate amounts of sulphates for their nourishment. Other optimum conditions for their growth have been found to be a temperature of 25 to 30°C and pH ranging from 5 to 9. The mechanism of anaerobic microbiological corrosion of iron is believed to occur through the following reactions:

Anodic solution of iron:

$$8H_2O \rightleftharpoons 8H^+ + 8OH^-$$

$$4Fe + 8H^+ \rightleftharpoons 4Fe^{2+} + 8H$$

Depolarization due to activity of bacteria:

$$H_2SO_4 + 8H \rightleftharpoons H_2S + 4H_2O$$

Corrosion products:

$$Fe^{2+} + H_2S \rightleftharpoons FeS + 2H^+$$

$$3Fe^{2+} + 6OH \rightleftharpoons 3Fe(OH)_2$$

The principal corrosion products are black iron sulphide and ferrous hydroxide. The corrosion is intense and localized.

(2) *Sulphur bacteria,* like *thiobacillus*, are mostly aerobic. Such bacteria oxidize sulphur present in their cells to yield sulphuric acid, which attacks the iron. They grow best under acid conditions of a pH from 0 to 1.

(3) *Iron and managanese micro-organisms* are aerobic micro-organisms and live by taking into their cells iron and manganese ion, which they digest in presence of oxygen forming insoluble hydrates of iron and manganese dioxide. The insoluble hydrates are then thrown out of the bodies of these micro-organisms. Iron bacteria grow in stagnant or running water at temperatures from 5°C to 40°C and pH between 4 to 10 and with a small amount of free dissolved oxygen.

(4) *Film-forming micro-organisms,* like bacteria, fungi, algae, diatoms, can form microbiological film on iron surface. Such films are capable of maintaining concentration gradients of dissolved salts, acids and gases on the iron surface, thereby leading to the formation of local biological concentration cells and consequent corrosion.

PREVENTION OF CORROSION

As we know that corrosion reactions are all surface reactions it means that protection of the surface of the metal should be the primary object to prevent its corrosion. The methods adopted for protection of metals can be classified into four main parts:

(I) Methods based on the proper disigning.

(II) Methods based on the treatment of metals.

(III) Methods based on the treatment of medium.

(IV) Methods based on the external influences.

(I) Methods Based on Proper Designing

A proper selection of metarial for any particular environment and proper design is best way of controlling corrosion. The design of the material should be such that "corrosion even if it occurs, is uniform and does not result in intense and localized corrosion". Important design principles are :

(i) "*Avoid the contact of dissimilar metals in the presence of a corroding solution*". *If this principle is not taken into consideration during designing, then corrosion is localized on the more active metal in the immediate vicinity of contact; while the other metal (less active) being protected.*

(ii) *When two dissimilar metals are to be in contact, the anodic material should have large area as possible; whereas the cathodic metal should have as much smaller area as possible.*

If anodic metal area is small and the cathodic metal is much larger in area (e.g., a steel pipe in a large copper tank), then cathodic area produces a large corrosion current, which is then localized in smaller area of anodic metal (e.g., iron in the above case) and causes rapid destruction.,

(iii) *If two dissimilar metals in contact have to be used, they should be as close as possible to each other in the electro-chemical series.*

(iv) *Whenever the direct joining of dissimilar metal is unavoidable, an insulating fitting may be applied in between them to avoid metal-metal electrical contact.*

(v) *The anodic metal should not be painted or coated when in contact with a dissimilar cathodic metal, because any break in coating would lead to it rapid localized corrosion.*

(II) Methods Based on the Treatment of Metals

This includes the following methods:

(a) Alloying of the metal : By alloying, corrosion can be decreased. This is successful in controlling corrosion in case of iron, nickel and copper as they have greater tendency to form alloys. Alloying of metals resists corrosion in two ways:

(i) *In some cases, homogeneity of the metal is increased which results in a decrease in corrosion.*

(ii) *In some other cases, oxide films are formed on the surface of the metals. An example is silicon-iron alloy called Durrion which is very resistant to acids due to the formation of silicon oxide film on its surface.*

(b) Removal of strains : The strains in the metals can be eliminated by heat treatment and annealing as these strains are focal points for collision reactions.

(c) Polishing the surface : By polishing the surface, the corrosion of metals can be decreased. This decrease in corrosion occurs due to

(i) *the removal of oxygen concentration cell and*

(ii) *decreasing the surface area because a smooth surface has a small area exposed to corrosive medium than the non-polished surface.*

(d) Producing insoluble oxide coating : In certain metals corrosion causes the formation of a tenacious film on the metal surface giving protection to a further deeper corrosion. For example metals like chromium, aluminium, nickel and others are capable of forming insoluble oxide films on their surfaces. These protective films can be produced by the artificial methods like:

(i) Rawer and Barf Process : A protective film Fe_3Cl_4 is produced on the surface of iron by heating it in steam at 900°C

(ii) Rffiungaton Process : In this process, the protective coating can be produced by placing the metal in a molten mixture of potassium or sodium nitrate and manganese dioxide.

There are several other processes. For example, protection of aluminium is brought about by electro-chemical surface oxidation.

(e) Electroplating : Noble and base metals are used for electroplating on any desired metal. When a noble metal (Au, Ag, W, Pt) is electroplated on the base metal, it protects the metal (base) by virtue of its noble character. Generally, tin plating and nickel plating are employed. As far as the covering is continuous, it affords protection.

Covering with metals by methods other than electron deposition : Recently, several methods have been developed to deposit one metal on another. Some of these are described below:

(i) Hot-dpping : This method is used for coating metal with films of metals having low melting points, such as zinc, tin, and lead. This process is carried out by dipping the cleaned metal in the hot solution of the metal. It is reported that during hot dipping an alloy is formed where the two metals meet. An interesting example is the galvanising of iron (*i.e.,* covering of iron surface by zinc metal) in which one ounce of zinc is enough to coat one sq. feet of the surface. The protective action of zinc is explained as below :

The thin coating of zinc in galvanised iron does not permit iron to have a direct contact with air and moisture and hence no corrosion (rusting) takes place. Even when the protective zinc coating is broken, iron is still protected because zinc occupies a position above iron in electrochemical series and hence, it is more active than iron. Hence in this case, it is the zinc which loses electrons (not iron) and then it passes into a solution as Zn^{2+} ions and the anodic area. Hydrogen is discharged from the surface of the iron in the cathodic area. Zinc is thus consumed in the protection of iron.

At area where the surface coating is not damaged, zinc is not corroding itself due to adhering of the layer of zinc oxids on it. Similarly, one can explain the protective action of tin. As tin is resistant to corrosion, a thin coating of tin over iron protects it from rusting. Tin coating, however, is not so durable as that of zinc in galvanised iron.

(ii) Metal spraying : A coating can be produced on metal surface by spray. This method is utilised for huge structures like bridges in which there is no need for dismantling them and metal coatings can be applied at any specific spot. Thus this process is only used when hot dipping is not possible.

(iii) Cementation : In this process, the metal to be deposited forms alloys with the surface of the metal on which deposition is taking place, *i.e.,* a surface alloy is formed with the depositing metal.

The articles to be treated are packed in powdered coating metal in a drum which can be rotated and which is heated at a temperature somewhat below the melting point (600°C) of the more fusible metal. Although this process is costly, it furnishes uniform thin films and is used in coating screen, bolts, nuts and other similar important articles.

(g) Covering with paints : Protective layers can be formed on the metal surface by paints, glass, oils, tars, bakelite, silica etc. However, paints and glass in the form of enamelling are generally used.

Red paint is ideal for iron surface. In paints there is generally primary coating which contains zinc chromate and red lead and is followed by secondary coating containing ferric oxide which gives mechanical protection. The only limitation of this method is the service behaviour of the protective coatings: (*i*) *The coating applied must be chemically inert to the environment under particular conditions of temperature and pressure*, (*ii*) *Moreover, they must prevent the penetration of the environment to the material, which they protect.*

(h) Making the metal cathodic by external e.m.f. : The e.m.f. of corrosion reactions is very small. When an external e.m.f. of greater magnitude and in the opposite direction is applied to metals, anodic metals will become cathodic and will not corrode. For this reason underground pipelines, which undergo intense corrosion, are made cathodic.

A potential of 6 to 10 volts is sufficient to prevent corrosion.

Methods Based on the Treatment of Medium

(a) Change of state of medium : If the state of medium is changed, the rate of corrosion is altered. It means that if the hydrogen ion concentration, conductivity or concentration of oxygen of the corroding medium is changed, the rate of corrosion is decreased to a great extent. For this reason, small amounts of alkali metal carbonates or hydroxides are added to change the hydrogen ion concentration of the corroding medium. Sometimes, potassium dichromate is added to the medium to produce protective oxide films on the metal.

(b) Corrosion Inhibitors : *There are certain substances which bring about chemical alteration of the metal surface so as to make the same corrosion proof. Such substances are called corrosion inhibitors.* A large number of organic compounds have been found to be effective but the best effective substance so far examined is quinoline. Potassium chromate is also used as a corrosion inhibitor. A chromate compound is generally added to the cooling water in radiators of cars to prevent corrosion.

According to a recent theory, due of *G.H. Cartladge,* the action of a corrosion inhibitor is to inactivate a metal surface chemically and protects it from rusting. It is important to remark that the effectiveness of an ion in inhibiting corrosion depends on the amount of the central positive charge. It is recommended that a very dilute solution of potassium pertechntity $KTcO_4$ may be used as a corrosion inhibitor. Unfortunately it could not be done so because its quantity available in the world is not sufficient for this purpose. However, it is expected that in the near future it will become a tool of research in corrosion and other spheres. Recently, tannic acid is used as an inhibitor. It is found to be very effective and is being under investigation for its use on a commercial basis. There are two types of inhibitors:

(i) *Anodic inhibitors (such as chromate, phosphates, tungstates or other ions of transition elements with a high oxygen content) are those that stifle the corrosion reaction occurring at the anode by forming a sparingly soluble compound with a newly produced metal ion. This compound is adsorbed on the metal surface forming a protective film or barrier, thereby reducing the corrosion rate. Although this type of control is effective, it may be dangerous since severe local attack can occur if certain areas are left unprotected by depletion of the inhibitor.*

(ii) *Cathodic inhibitors are of two types: In acidic solutions, the main cathodic reaction is evolution of hydrogen.*

$$2H^+ + 2e^- \longrightarrow H_2$$

Consequently, corrosion may be reduced either *by slowing down the diffusion of hydrated H^+ ions to the cathode and by increasing the over potential of hydrogen evolution.* The diffusion of H^+ ion is considerably decreased by organic inhibitors like amines, mercaptans, heterocyclic nitrogen compounds, substituted ureas and thioureas, heavy metal soaps, which are capable of being adsorbed at the metal surface. On the other hand, antimony and arsenic oxide or salt like sodium arsenite are used as inhibitors, because they deposit adherent film of metallic arsenic or antimony at the cathodic areas, thereby increasing considerably the hydrogen overvoltage.

(IV) Methods Based on the External Influence : Methods should be adopted to the following processes to check corrosion due to the external influence:

(i) *Painting the metal with light proof paint: This is done to keep off light from the surface of the metal.*

(ii) *Detection and removal of stray currents.*

(iii) *Prevention of bacterial growth : Suitable antiseptics and bactericides are being generally added to the corroding medium to check corrosion.*

From the above methods, it is evident that corrosion of the metals can be prevented in an effective manner by the methods based on their treatment.

4

Coordination Complexes of Metal Alkoxides

INTRODUCTION

Metal alkoxide possess little tendency to form co-ordination compounds with donor molecules. In *1962, Bradley* and *Bains* prepared a number of co-ordination complexes of aluminium, titanium, zir-conium, and tantalum alkoxides with ethylenediamine and hydrazine.

In 1962, **Mehrotra** and **srivastava** prepared a number of pyridine and diethylamine complexes of triaryl borates. Gilman reported that uranium penta-2, 2, 2, trifluoro ethoxide formed stable addition compounds with ammonia or aliphatic amines.

In 1965, **V. D. Gupta** reported that he has isolated a disubstituted product $Sn(OR)_2$ (en) in the reaction of diamine with tin (IV) alkoxides in equimolar ratio.

Some examples of the addition compounds of the alkoxides of lower aliphatic alcohols are $Ti(OBu)_4.Bu^tOH$, $Zr(OPr^i)_4.Pr^i(OH)$, $Zr(OPr^i)_4.C_5H_5N$, $Ce(OPr^i)_4.Pr^i(OH)$ and $Ce(OPr^i)_4.C_5H_5N$.

Paul and co-workers have recently reported the formation of addition compounds of alkoxides with nitrogen and oxygen donor molecules.

STRUCTURE OF METAL ALKOXIDES

Structures of only few alkoxides have been established with certainty. **Wheatly** (1959) studied the X-ray diffraction patterns of lithium methoxide. He reported that lithium methoxide has a two dimensional layer structure analogous to that of lithium hydroxide. The Li—O bond length is 1.90Å and O—CH_3 bond length is 1.42Å. Each oxygen atom is bonded to four lithium atoms and a methyl group in the form of square pyramid with oxygen at the centre and methyl at the apex. The bonding is said to be 'electron deficient' and 'delocalised' since the CH_3O– oxygen possesses only three pairs of electrons for co-ordination to four lithium ions.

Weiss elucidated double layer and polymer structure of potassium methoxide. From conductance measurements of sodium alkoxide, he showed that NaOMe, NaOEt and $NaOpr^i$ are ionic in the decreasing order of conductance whereas $NaOBu^i$ is covalent owing to its non conductance.

Mehrotra reported that the freshly distilled aluminium isopropoxide was irimer . But *Fieggen* (1970) reinvestigated the structure of isopropoxide and reported that the low melting form of aluminium isopropoxide was trimeric.

In the tetrameric structure of aluminium isopoproxide, the central aluminium atom has a co-ordination number of six whereas each of three surrounding aluminium atoms has a co-ordination number of four.

Caughian and coworkers (1951) reported that $Ti(OEt)_4$ and $Ti(OMe)_4$ were tetrameric. But Mehrotra and Verma reported that they were tetrameric. J.A. Ibres (1963) reported the structure of $Ti(OEt)_4$ which was based on a group of four octahedra TiO_6 which shared edges so that four titanium atoms remained in one plane and occupied the corners of a rhomb and formed a compact unit. The sturcture was confirmed by Bradley and Holloway by N. M. R. studies.

DOUBLE ALKOXIDES

Meerwein and *Bersin* have reported that alcoholic solutions of the alkoxides of aluminium, iron (III), titanium (IV), tin (IV) and zirconium (IV) are acidic in nature and these can be titrated to a sharp end-point with an alkali-metal alkoxide using thymolphthalein as an indicator. The reaction can be put as:

$$\underset{\text{Base}}{\text{NaOEt}} + \underset{\text{Acid}}{\text{Al(OEt)}_3} \rightarrow \underset{\text{Salt}}{\text{Na[Al(OEt)}_4\text{]}}$$

No evidence is available which could confirm the nature of compounds like $Na[al(OEt)_4]$ as ionic salts. However, physical properties such as solubility in organic solvents and volatilty indicate that these compounds are not ionic but covalent and can be best regarded as double alkoxides.

A study of complexes formed by aluminium isopropoxide with ethylene diamine has been carried out recenty by NMR technique.

POLYMERIC METAL ALKOXIDES

A striking difference between the aliphatic orthoesters [*e.g.*, $B(OR)_3$, $C(OR)_4$, $Si(OR)_4$] and the metal alkoxides $M(OR)_x$ is that the orthoesters are simple monomeric compounds, while the lower aliphatic alkoxides of the metals are polymeric. The polymeric nature of the metal alkoxides is ascribed to intermolecular coordination between the metal and the oxygen of a neighbouring alkoxide group. Such compounds are in fact co-ordination polymers $[M(OR)_x]_n$ and it is of interest to know whether there exists any relationship between the degree of polymerization n, the primary valency of the metal x, and the higher coordination number and stereo-chemistry of the metal. It follows that an increase in coordination number by unity could suffice to bring about the formation of an infinite linear polymer.

Table 1 : Degrees of polymerization of metal normal alkoxides.

Metal alkoxide	*Degree of Polymerization*	*Metal alkoxide*	*Degree of Polymerization*
TlOR	4	$Th(OR)_4$	6–7
$Al(OR)_3$	4	$V(OR)_4$	2–3
$Fe(OR)_3$	3	$Nb(OR)_5$	2
$Ti(OR)_4$	2–3	$Ta(OR)_5$	2
$Zr(OR)_4$	3–4	$U(OR)_4$	2–3
$Ce(OR)_4$	4–5		

STRUCTURAL ASPECTS OF METALS ALKOXIDES

One of the first structural predictions in this field was that made by *Sidgwick* and *Sutton* for tetrameric thallous alkoxides. Preliminary reports on the X-ray diffraction of thallous alkoxide have been found to be essentially in agreement with this structure. It seems probable that the orbital used by oxygen and thallium get directed tetrahedrally (sp- hybrids) with the fourth orbital of each thallium occupied by the lone pair of electrons.

Robinson and Peak revealed a structure for the aluminium alkoxides based on the tetrahedral coordination of the aluminium. However, it is not all clear why a cyclic tetramer should be a preferred structure.

In 1951, *Caughlan* and co-workers reported on the cryscopically determined molecular weights of Ti(OEt)4, Ti(OPr")4, Ti(OPr')4, Ti(OBu").i and TiCl(OEt)3 in benzene. In each case, the number of degree of polymerization got increased with increasing concentration and reached a maximum of trimeric at a mole fraction of about 0.01. It was therefore postulatad that trimeric unit Ti.i(OR)i2 of structural significance and two possible structures were proposed based on the octahedral 6-coordination of titanium following the example of rutile (TiOz). An interesting feature of the polymeric metal alkoxides is the tendency of the metal to expand its coordination number by the polymerization process rather than by coordination with another ligand.

Also a common feature of all the structures mentioned in the previous section in the presence of four membered rings involving two metals and two oxygens has been depicted as follows :

Evidently the bridge structure has enhanced stability, which must presumably involve an entropy effect.

Industrial Applications

Main applications are :

1. Metal alkoxides can be used to accelerate the drying of paints or printing inks. For example, titanium or zirconium alkoxides are added to bring about the rapid drying of printing inks.
2. Less reactive titanium alkoxides are generally added to paints and lacquers to increase the adhesive power of paints and lacquers.
3. Aluminium isopropoxide is added to a paint to remove water from its pigment and to decrease its wrinkling tendency and to increase water resistance.
4. $Al(OR)_3$ is used as a lubricant.
5. Metal alkoxides are used as effective catalysts in the conversion of an aldehyde to an ester, conversion of ketones to secondary alcohols, etc. For example, sodium methylate is used as an effective catalyst in the condensation reactions.
6. Zirconium alkoxides have been used to impart water repellency to textiles. Similarly, a solution having 2.5% silicone and 0.2% aluminium ethoxide in benzene is being used to make textiles fibres water repellent.

7. A mixture of titanium butoxide and aluminium are used for the coating for optimum heat and corrosion resistance.

NITROSYLS

Nitrosyls are the compounds in which the nitrogen of the nitrosyl group is directly bonded to the atoms or ions.

or

The compounds containing nitric oxide group are called nitrosyl compounds.

Nitric oxide molecule as such is an odd electron molecule having an unpaired electron. On account of its structure, it readily unites with other elements by direct addition to form nitrosyl compounds. Nitric oxide might form nitrosyl compounds by the following three ways:

(i) *A positive ion,* NO^+, *is formed due to the loss of an electron which then combines with atom or molecule. It may have the following structure:*

$$(: P : : : O :)^+ \text{ or } (: P = O :)^+$$

(ii) *A negative ion* NO^- *is formed due to the gain of an electron from some electropositive metal and it has probably the structure:*

$$(:\ddot{N}:::O:)^+ \text{ or } (:\ddot{N}=\ddot{O}:)^-$$

(iii) *Nitric oxide may act as a co–ordinating group through the donation of an electron pair. Such behaviour might involve neutral molecule or* NO^+ *or* NO^- *group.*

CLASSIFICATION OF COMPOUNDS CONTAINING THE NO^+ GROUP

In terms of the above consideraions, the nitrosyl compounds are classified as given below:

Examples are:

(i) $NOHSO_4$ and $(NO)_2S\ 2O_7$,

(ii) Nitrosyl Halides (NOX, where X = F, Cl, Br),

(iii) $NOClO_4$, $NOBF_4$

(iv) $(NO)_2SeO_4)$ $NOHSeO_4$,

(v) NOSCN,

(vi) $NO[Cr(NH_3)_2\ (SCN)_4]$,

(vii) $NOSbCl_6$ $NOAsF_6$, $(NO)_2$ $PtCl_6$, $NOFeCl_4$ and $NONO_3$.

In the above compounds, Angus and Leckie confirmed the presence of NO^+ group by the following observations:

(i) Ionization potential of nitric oxide ($NO \rightarrow NO^+ + e^-$) is 9.5 eV which is lower than that for nitrogen ($N_2 \rightarrow N_2^+ + e^-$) (~ 16 eV) and oxygen ($O_2 \rightarrow O_2^+ + e^-$) (~ 15 eV). This makes removal of an electron easy.

(ii) As the ionisation potential of nitric oxide is low, it can lose one electron easily to form NO^+ ion.

(iii) Both nitrosyl hydrogen sulphate $NOHSO_4$ and nitrosyl perchlorate $NOClO_4$ exhibit Roman displacements which clearly confirm the presence of NO^+ group.

(iv) When electrolysis of a solution of nitrosyl hydrogen sulphate in cone. H_2SO_4 is carried out by using an iron cathode, a brown colour of $Fe(NO)^{2+}$ is developed around the cathode which indicates the migration of the nitrosyl groups to the cathode.

(v) When solutions of nitrosyl hydrogen sulphate and perchlorate are prepared by dissolving in nitromethane, they form conducting solutions and behave as binary electrolytes.

(vi) Salts of nitrosyl sulphuric acid could not be prepared, confirming that nitrosyl group is positive one, *i.e.*, NO^+.

(vii) *Cryoscopic* and *Roman data* confirm clearly that dissolution of the oxides N_2O_3 and N_2O_4 in sulphuric acid gives the NO^+ group in terms of the following equations:

$$N_2O_3 + 3H_2SO_4 \longrightarrow 2NO^+ + 3HSO_4^- + H_3O^+$$

and $$N_2O_4 + 3H_2SO_4 \longrightarrow NO^+ + NO_2^+ + 3HSO_4^- + H_3O^+$$

Preparation of Compounds Containing NO^+ Group

[A] Nitrosyl Halides: These may be prepared by the direct combination of nitric oxide with halogens. However, we shall consider these one by one:

(a) Nitrosyl Fluoride, NOF

Preparation

(i) *It may be prepared by the direct combination of nitric oxide and fluorine in copper tube.*

$$2NO + F_2 \longrightarrow 2NOF$$

(ii) *It may also be prepared by heating silver fluoride with nitrosyl chloride in a platinum vessel.*

$$NOCl + AgF \longrightarrow NOF + AgCl$$

(iii) *It may also be obtained by heating sodium fluoride with nitrosyl fluoroborate.*

$$NOBF_4 + NaF \longrightarrow NOF + NaBF_4$$

Properties

It is a colourless reactive gas (f.p. – 132.5°C b.p. – 60°C). It reacts slowly with oxygen difluoride to yield oxygen and nitrogen trifluoride.

$$NOF + OF_2 \longrightarrow O_2 + NF_3$$

It reacts with water and alcohol as follows:

$$NOF + H_2O \longrightarrow HNO_2 + HF$$

$$NOF + ROH \longrightarrow RONO + HF$$

It attacks glass. It forms addition compound with liquid SO_2 as $NOF.SO_2$.

(b) Nitrosyl Chloride, NOCl

Preparation

It is prepared by any of the following methods :

(i) By *the action of* PCl_5 *on* KNO_2

$$PCl_5 + KNO_2 \rightarrow POCl_3 + KCl + NOCl$$

(ii) *By decomposition of ferric chloride with nitric oxide at high temperature.*

$$FeCl_3 + NO \rightarrow FeCl_2 + NOCl$$

(iii) *By warming* $NOHSO_4$ *with dry* NaCl *below* 80°C.

$$NOHSO_4 + NaCl \rightarrow NOCl + NaHSO_4$$

(iv) *By passing dry nitric oxide into liquid chlorine.*

$$NO + Cl_2 \rightarrow NOCl_2$$

$$NOCl_2 + NO \rightarrow 2NOCl.$$

Properties

It is yellow gas at ordinary temperature with suffocating odour (f.p. – 61.5°, b.p. – 55°C). On cooling it condenses to a yellowish red liquid. It decomposes at 170°C.

$$2NOCl \rightarrow 2NO + Cl_2.$$

It reacts with water and alcohol as follows:

$$NOCl + H_2O \rightarrow HCl + HNO_2$$

$$NOCl + ROH \rightarrow RONO + HCl$$

It reacts with hydrogen sulphide gas as follows:

$$NOCl + 3H_2S \rightarrow NH_4Cl + H_2O + 3S$$

It forms addition compounds with unsaturated organic compounds and with many metallic chlorides.

$$>C=C< + NOCl \rightarrow \underset{NO}{>C}-\underset{Cl}{C<}$$

$$CuCl_2 + NOCl \rightarrow CuCl_2NOCl$$

$$SnCl_4 + 2NOCl \rightarrow (NO^+)_2\,(SnCl_6)^{2-} \quad SbCl_5 + NOCl$$

$$\rightarrow NO^+\,SbCl_6^-$$

Uses

It is known as *Tilden* regent and lines use for knowing the presence and number of double bonds.

(c) Nitrosyl Bromide, NOBr

Preparation

It may be prepared by passing nitric oxide into bromine at –15°C.

$$2NO + Br_2 \rightarrow 2NOBr$$

Properties : It is a dark brown liquid (m.p. – 55.5°C). It is less stable than NOF and NOCl. It decomposes even at room temperature.

Structure of Nitrosyl Halides

In nitrosyl halides, the N – Cl and N – Br bond lengths are greater than the sum of covalent bond radii of 1.73 Å (N – Cl) and 1.88 Å (N – Br) but less than ionic bonds *e.g.*, 2.64 Å for (N – Cl) based upon 1.40 Å as the radius of NO^+. As the (N – O) bonds are intermediate in lengths between double and triple bonds, it means that the following resonance structures characterise the nitrosyl halides:

The above structures have been confirmed by comparing the measured dipole moments with the calculated values of purely covalent nitrosyl compounds. They are in full agreement, indicating that both the chlorine and bromine are about 50% ionic. The bond lengths, bond angles and structure of nitrosyl halides etc. are summarised as follows:

Table 2 : Bond lengths, bond angles and structure of NOX

Parameter	*NOF*	*NOCl*	*NOBr*
N – X distance (Å)	1.52	1.97	2.13
N – O distance (Å)	1.13	1.14 ± 0.02	1.15 ± 0.04
X – M – O angle	110°	113°	117°
Structure	Bent	Bent	Bent
Dipole moment	—	2.19 D	1.90 D
Ionic character (δ)	> 50	50	50

According to orbital theory, it may be considered that nitrogen atom exhibits *sp* hydridization, forming c bonds to halogen and oxygen atom and having an unshaired pair, and then a π bond, using its p–orbitals to oxygen.

[B] Nitrosyl Sulphuric Acid or Hydrogen Sulphate

Preparation

(i) *By passing nitrogen (III) or (IV) oxide into conc.* H_2SO_4.

(ii) *By the action of nitrosyl chloride on conc.* H_2SO_4

$$NOCl + H_2SO \rightarrow NOHSO_4 + HCl$$

(iii) *As an intermediate in the Chamber process for the manufacture of* H_2SO_4.

Properties

It is a colourless, hygroscopic crystalline solid (m.p. 75°C). It forms conducting solution in nitromethane. It decomposes in moisture as:

$$NOHSO_4 + H_2O \rightarrow H_2SO_4 + HNO_2$$

It reacts with sodium chloride to form nitrosyl chloride.

$$NOHSO_4 + NaCl \rightarrow NaHSO_4 + NOCl$$

[C] Nitrosyl Perchlorate, $NOClO_4$

Preparation

It is prepared by passing a gaseous mixture of NO and NO_2 in an aqueous solution of sodium perchlorate and concentrated sulphuric acid.

$$NO_2 + NO + NaClO_4 + H_2SO_4 \rightarrow NOClO_4 + NaHSO_4 + HNO_2$$

Properties

It is colourless crystalling solid. It decomposes on heating.

$$2NOClO_4 \rightarrow N_2O_4 + Cl_2 + 3O_2$$

[D] Preparation of $NOBF_4$ $NOReO_4$ $NOHSeO_4$, etc: All the compounds can be prepared by passing nitrogen (II, III or IV) oxide into the appropriate acids.

[E] Preparation of $NOSbCl_6$, $(NO)_2SnCl_6$: These compounds can be prepared by the addition of nitrosyl halide to an appropriate halide.

It is important to remark that compounds containing the NO^+ groups are generally susceptible to hydrolysis and must be under anhydrous conditions.

COMPOUNDS CONTAINING NO^- GROUPS

The only example of this type is sodium nitrosyl.

Sodium nitrosyl is prepared by the action of dry nitric oxide on sodium in liquid ammonia.

$$Na + NO \rightarrow NaNO$$

Sodium nitrosyl is a white substance which is unstable. This compound is diamagnetic which clearly confirms the presence of NO^- group.

It is a white and unstable solid. Magnetic properties suggest the presence of the NO^- group.

It is decomposed by water with the evolution of nitrous oxide and the formation of sodium hydroxide, sodium hyponitrite and traces of sodium nitrate.

Its probable structure is:

$$Na^+ \ [\ddot{N} = \ddot{O}:]^-$$

The bond angle is around 125° in this molecule.

COORDINATION COMPOUNDS OR METAL NITROSYLS

These are the compounds in which the nitrogen of the nitrosyl group is directly bonded to the metal atoms or ions. These compounds are also known as metal nitrosyls

Metal nitrosyls occur mostly as mixed ligand complexes in conjugation with other π-bonding low oxidation state stabilising ligands.

The properties and compositions of these compounds vary widely, therefore, it is best to consider them under the broad heading of co-ordination compounds.

In metal nitrosyls or co–ordination compounds, the nitric oxide acts as a co–ordinating group through the donation of an electron pair to the metal atom or ion. Such behaviour might involve the neutral molecule or the NO^+ or NO^- group.

(1) Co–ordination compounds containing neutral nitric oxide molecule: The co-ordination by neutral nitrosyl group is unlikely due to the lack of experimental evidences. Some of these are considered below:

(i) Iron penta carbonyl on reaction with nitric oxide forms volatile nitrosyl carbonyl $Fe(NO)_2(CO)_2$. The similarities of this compound with metallic carbonyl $[Fe(CO)_5]$ suggest co-ordination by neutral nitric oxide molecules. As the neutral nitric oxide molecule contains one unpaired electron, it clearly suggests that $Fe(NO)_2(CO)_2$ must be paramagnetic but actually it is diamagnetic. Thus, this example clearly indicates that co–ordination by neutral nitric oxide molecule is unlikely.

(ii) If the neutral nitric oxide molecules were involved, they could be replaced by neutral molecules without any change in the compound type. When nitrosyl carbonyls are treated with pyridine, it replaces carbonyl group without affecting the nitrosyl groups. It confirms that coordination by neutral nitrosyl group is unlikely.

(2) Co–ordination compounds containing NO^- group: In general, this type of co–ordination compounds is also unlikely because only positive evidence for this is $[(Co(NH_3)_5NO]^{2+}$ where cobalt (II) apparently transfers an electron to the NO group. There are two isomers of $[Co(NH_3)_5(NO)]^{2+}$ red and black.

The red isomer is dimeric, and the black isomer is monomeric. Actually, this structure is reasonable in some ways but puzzling in others. This is a diamagnetic pink compound containing NO^- group.

The compound $[Co(NH_3)_5NO]^{2+}$ is analogous to the halopentammines ruthenium complexes, etc.

(3) Co–ordination compounds containing NO^+ group: The evidence for the co–ordination compounds containing NO^+ group is more convincing. This nitric oxide cation NO^+ is similar to the neutral carbon monoxide molecule and the cyanide ion.

$$:N:::O:^+ \quad :C:::O: \quad :C:::N:^-$$

Corresponding similarities have been reported among compounds containing these materials. Examples are:

$$K_2[Fe(CN)_5NO],\ K_3[Fe(CN)_5CO] \text{ and } K_4[Fe(CN)_6]$$

The co–ordinated nitrosyl compounds may be classified as:

(A) Nitrosyl carbonyls and related compounds: The examples of this type are the cobalt and iron compounds $Co(NO)(CO)_3$ and $Fe(NO)_2(CO)_2$. Nitrosyl carbonyls resemble the carbonyls chemically.

As CO donates two electrons to the metal atom in the formation of metallic carbonyls while NO molecule donates three electrons to the metal atom in the formation of nitrosyls, there is possibility that two NO molecules can replace three CO groups from metallic carbonyls to form nitrosyl carbonyls. Examples of pairs of compounds so related are given as follows (oxidation state of the metal atom is shown in parentheses):

$$Fe(CO)_5\ (Fe = 0)\ Fe(CO)_2(NO)_2\ (Fe = -2)$$

$$Mn(CO)_4(NO)[Mn = -1]\ Mn(CO)(NO)_3[Mn = -3]$$

$$Co(CO)_3(NO)[Co = -1]\ Co(NO)_2[Co = -3]$$

Preparation

(i) The nitrosyl carbonyl compounds like $Co(CO)_3$ NO and Fe(NO)2$(CO)_2$ can be prepared by the action of nitric oxide on polynuclear carbonyls of cobalt and iron.

$$Co_2(CO)_8 + 3NO \rightarrow 2Co(CO)_3NO + 2CO$$

$$[Fe(CO)_4]_3 + 6NO \rightarrow 3Fe(NO)_2(CO)_2 + Fe(CO)_5 + CO$$

(ii) Cobalt nitrosyl carbonyl may be prepared by the action of nitric oxide on solutions containing the $[Co(CO)_4^-$ ion

(iii) Cobalt nitrosyl carbonyl may also be prepared by treating an alkaline suspension of cobalt (II) cyanide first with carbon monoxide and then with nitric oxide.

(iv) When dry nitric oxide is passed through trimeric iron tetracarbonyl $[Fe(CO)_4]_3$, it gives a mixture of iron nitrosyl carbonyl and iron pentacarbonyl. Related nitrosyls of iron $[Fe(NO)_4]$ *and ruthenium* $Ru(NO)_4$ *and* $Cu(NO)_5$ *have been reported to be prepared by the action of nitric oxide on carbonyls* $Fe(CO)_5$ *and* $Ru_2(CO)_5$ *respectively.*

$$Fe(CO)_5 + 4NO \rightarrow Fe(NO)_4 + 5CO$$

These compounds have not been studied completely and structural data are lacking.

Properties

(i) ***Substitution reactions***: *In metal carbonyl nitrosyls,* NO^+ *ions are more firmly attached with the metal ion than the CO groups. It is for this reason that when metal carbonyl nitrosyls get treated with ligands like* PR_3, CNR, *phen etc., it is only* CO *groups that are replaced by these ligands, e.g.,*

$$Fe(CO)_2(NO)_2 + 2L(L = PR_3, CNR) \rightarrow Fe(L)_2(NO)_2 + 2CO$$

$$Fe(CO)_2(NO)_2 + phen \rightarrow Fe(phen)\,(NO)_2 + 2CO$$

(ii) ***Action of halogens***: *Many metal carbonyl nitrosyls, when treated with halogens, get converted into metal nitrosyl halides, e.g.,*

$$2[Fe(CO)_2(NO)_2] + I_2 \rightarrow [Fe(NO)_2I]_2 + 4CO$$

In the nitrosyl carbonyls the NO^+ ion is more firmly attached to the metal ion that the CO group which follows from the fact that CO group present in nitrosyl carbonyls can be easily replaced by other K–bonding ligands such as PR_3, CNR, phen, etc. to form other complexes. For example, the reaction of these ligands with $Fe(CO)_2(NO)_2$ yields $[Fe(NO)_2(PR_3)_2]$, $[Fe(NO)_2(CNR)_2]$ and $[Fe(NO)_2\,(phen)]$ respectively.

(B) Nitrosyl compounds of type $(M^1(NO)_X\ A_y$: Main types of nitrosyls of monopositive metals are:

(1) ***Metal nitrosyl hydroxides***: *The exampiles of this are* : *Ni(NO)OH and its alcohol derivatives like Ni(NO)(OR)OH and Ni(OH)(OR).x(ROH)* H_2O [R = CH_3 (x = 1), C_2H_5 (x = 0)]. *Metal nitrosyl hydroxides are strong reducing compounds and have been prepared by the action of nitric oxide on nickel tetracarbonyl in the presence of water or water and alcohol.*

(2) ***Metal nitrosyl halides***: *Examples are*:

(a) $Fe_2(NO)_4I_3$ $Fe(NO)_2I$, $Fe(NO)I$, $Fe(NO)_3Cl$

(b) $Co(NO)_2X$, where X = Cl, Br, I

(c) Ni (NO)X where X = Cl, Br, I

Preparation

(a) Volatile diamagnetic trinitrosyl iron halides Fe $(NO)_3X$, may be prepared by the direct action of nitric oxide on iron carbonyl halides in the presence of finely divided iron as a halogen acceptor. These readily lose nitric oxide, yielding the corresponding dinitrosyl halides. Similarly, nitrosyl halides, *i.e.,* $Co(NO)_2X$ and Ni(NO)X may be prepared.

(b) Nitrosyl halides of cobalt and nickel can be prepared by the action of nitric oxide on metallic in the presence of a suitable halogen acceptor.

$$CoX_2 + Co + 4NO \longrightarrow 2[Co(NO)_2X]$$

$$4NiI_2 + 2Zn + 8NO \longrightarrow 2[Ni(NO)I]_4 + 2ZnI_2$$

$$SnI_4 + 2Co + 4NO \longrightarrow 2Co(NO)_2I + SnI_2$$

Ease of the formation of these compounds increases in the sequence Ni < Co < Fe and X = Cl < Br < I. Thus, $Fe(NO)_2I$ is the most stable compound and is readily formed by the direct action of nitric oxide on cobalt iodide in the presence of metallic iron.

$$CoI_2 + 2Fe + 4NO \longrightarrow Co + [Fe(NO)_2I]_2$$

(c) Nitrosyl halides may also be prepared by the action of halogen on nitrosyl carbonyls.

$$2[Fe(CO)_2(NO)_2] + I_2 \longrightarrow [Fe(NO)_2I]_2 + 4CO$$

The $M(NO)_2Cl$ (M = Mo or W) may also be prepared as follows.

$$M\,(NO)_2Cl_2 + 2\overset{+}{NO}\overset{-}{Cl} \xrightarrow[20°C]{CH_2Cl_2} M(NO)_2Cl_2 + 6CO$$

Properties of Metal Nitrosyl Halides

(i) *Metal nitrosyl halides react with other ligands to form mono–nuclear complexs, e.g.,*

$$[Fe(NO)_2X]_2 + 2L \rightarrow 2\,[Fe(NO)_2XL]$$

(ii) *Iron nitrosyl halide,* $[Fe(NO)_2I]_2$ *reacts with* K_2S *and* CH_3Cl *to form dark red compounds which have the composition,*

$K_2[Fe(NO)_2S]_2$ and $[Fe(NO)2(SCH_3)_2$ *and are called Roussin's salts. In these compounds Fe is in – 1 oxidation state.* $[Fe(NO)_2I]_2 \xrightarrow[-2KI]{+K_2S} K_2[Fe(NO)_2S]_2$

$$\xrightarrow[-2KCl]{+CH_3Cl} [Fe(NO)_2(SCH_3)]_2$$

(3) ***Metal nitrosyl thio-compounds***: *The true structures of most of the metal nitrosyl thiocompounds have not been confirmed. These compounds appear to have unipositive metals but the exact oxidation states of the metals have never been determined with certainty. Iron forms the compound* $Fe(NO)_2SA$, *where A may be hydrogen or a metal, sulphonic acid groups, or an alkyl or aryl group.*

Best examples are Roussin's red salts, M[Fe $(NO)_2S]$ where M = Na^+. K^+, and NH_4^+. These are unstable compounds which have been prepared by the action of nitric oxide on freshly precipitated iron (II) sulphide.

Other examples are:

$M[Fe(NO)SSO_3[Fe[(NO)_2SR]$ [R = ethyl or phenyl group etc].

Iron nitrosyl halides react with alkali sulphide K_2S and alkyl halides (CH_3Cl, C_2H_5Cl) to form the dark red compounds which are known as Roussin's salts.

$$[Fe(NO)_2I]_2 \xrightarrow[-KI]{K_2S} K_2[Fe(NO)_2S]_2 \xrightarrow[-KCl]{2CH_3Cl} [Fe(NO)_2(SCH_3)]_2$$

(C_5H_5) $Cr(NO)_2$ Cl may be prepared as follows:

$$CrCl_3 + Na^+(C_5H_5)^- \xrightarrow{NO} [(C_5H_5)\ Cr(NO)_2Cl]$$

This complex may be converted into its dimer as follows:

$$2[(C_5H_5)Cr(NO)_2Cl] \xrightarrow{NaBH_4} \underset{\text{Dimer}}{[(C_5H_5)Cr(NO)_2Cl]_2}$$

(C) Nitrosyl compounds of the type M $(NO)_xA_y$: Crystalline compounds $Fe(NO)HPO_4$ and $Fe(NO)SeO_4$. H_2O have been prepared. Iron (II) nitrosyl complexes are normally brown, but green and red species have also been prepared.

Nitroso ferrous sulphate, $FeSO_4$.NO or $[Fe^+(NO^+)]SO_4$: When, to the aqueous solution of a metallic nitrate (say $NaNO_3$) is added freshly prepared solution of $FeSO_4$ and a few drops of cone. HNO_3 along the sides of the test tube, a brown ring of nitroso ferrous sulphate, $[Fe^+(NO^+)]SO_4$ is formed at the junction of the two liquids in the test tube. The formation of nitroso ferrous sulphate takes place through the following equation:

(a) $6NaNO_3 + H_2SO_4 \rightarrow NaHSO_3 + HNO_3$

$[\text{or } NO_3^- + H^+ \rightarrow HNO_3]$

(b) $6FeSO_4 + 2HNO_3 + 3H_2SO_4 \rightarrow 3Fe_2(SO_4)_3 + 2NO + 4H_2O$

$[\text{or } 3Fe^{2+} + NO_3^- + 4H^+ \rightarrow 3Fe^{3+} + NO + 2H_2O]$

(c) $FeSO_4 + NO\ FeSO_4 . NO$ *or* $[Fe^+(NO^+)]^{2+}SO_4^{2-}$

$[\text{or } Fe^{2+} + NO \rightarrow [Fe^+(NO^+)]^{2+}]$

In aqueous solution $[Fe^+(NO^+)]^{2+}$ ion is better expressed as $[Fe(NO)(H_2O)_5]^{2+}$. It is a paramagnetic substance which is corresponding to the presence of three unpaired electrons, since solution magnetic measurements give 3.90 B.M. as the value of its magnetic moment. This value supports the fact that Fe is in +1 oxidation state in this complex ion *i.e.,* it is a high spin compled of Fe (+1) ($3d^7$ system) with NO^+. The complex has N – O stretching frequency at 1745 cm^{-1} which reveals the presence of strong π bonding and the intense brown colour strongly suggests Fe^+ –NO^+ charge transfer.

The formation of $[Fe^+NO^+)]^{2+}$ ion has been used in the detection of NO_3 ion in a given inorganic salt.

(D) Nitrosyl compounds of the type M (NO)xAy: The only known examples of this type are:

$Fe_2(NO)_2(SO_4)_3$, $B(NO)F_3$, $B(NO)Cl_3$ and $Ru(NO)(R_2NCS_2)_3$ where R = CH_3, C_2H_5, etc. Only these compounds have been prepared.

(E) Nitrosyl derivatives containing groups of the type [MA(NO)]$^{n-}$ This type includes compounds yielding the ions such as:

$[Mn(CN)_5NO]^{5-}$, $[Fe(CN)_5NO]^{2-}$, $[Mo(CN)_5NO]^{4-}$

$[Ru(CN)_5NO]^{3-}$ $[Co(NH_3)_5NO]^{2+}$, $[Co(NH_3)_5NO]^{2+}$ $[OsCl_5(NO)]^{2-}$ etc.

Iron compounds $[M(CNHNO)]^{n-}$ have been prepared by the action of nitric oxide on hexacyano ferrate (II) salts. $[Mn(CN)_5NO]^{3-}$ is obtained by saturating manganese (II) acetate solution containing cyanide with nitric oxide.

$$Na_4[Fe(CN)_6] + NO \rightarrow Na_2[Fe(CN)_5NO] + 2NaCN$$

Sodium nitroprusside, $Na_2[Fe^{2+}(CN)_5(NO^+)]$ is one of the important compounds of this type. It may be prepared by the reaction of sodium nitrite on sodium ferrocyanide.

$$Na_4[Fe^{2+}(CN)_6] + NaNO_2 + H_2O \longrightarrow Na_2[Fe^{2+}(CN)_5(NO^+)] + 2NaOH + NaCN$$

(ii) It may also be prepared by passing nitric oxide (NO) into acidified solution of $Na_4[Fe(CN)_6]$.

$$2Na_4[Fe(CN)_6] + H_2SO_4 + 3NO \longrightarrow 2Na_2[Fe(NO)(CN)_5] + 2NaCN + Na_2SO_4 + 1/2\ N_2 + H_2O$$

Properties

(i) *Sodium nitroprusside forms beautiful ruby red rhombic crystals which are soluble in water.*

(ii) *If freshly prepared sodium nitroprusside is added to a solution having sulphide ion (i.e.,* Na_2S *but not* H_2S*), a purple or violet colour is produced. The production of this colour is ascribed to the formation of* $Na_4[Fe^{2+}(CN)_5(NO^+)(S)]$. *The production of this purple or violet colour is used to confirm the presence of* S^- *ion in a given mixture.*

$$Na_2S + Na_2[Fe^{2+}(CN)_5(NO^+)] \longrightarrow Na_4[Fe^{2+}(CN)_5(NO^+)(S^{2-})]$$

(Violet or purple colour)

(iii) *Alkali sulphites give a rose red colour because of the formation of* $Na_4[Fe(CN)_5(NO)(SO_3)]$. *This reaction can be used to distinguish sulphites from thiosulphates which do not show this reaction.*

(iv) *With silver nitrate a flesh coloured* $Ag_2[Fe(CN)_5(NO)]$ *is*

$$2AgNO_3 + Na_2[Fe(CN)_5(NO)] \longrightarrow Ag_2[Fe(CN)_5(NO)] + 2NaNO_3$$

(v) *Aldehydes and ketones having* $CH_3 - CH_3 - \overset{\overset{O}{|}}{C} - R - R$ *group yield deep red colour with sodium nitro-prusside and excess* of NaOH.

(vi) *It gets converted into sodium ferrocyanide,* $Na_4[Fe(CN)_6]$ *on treatment with an alkali.* $6Na_2[Fe(CN)5(NO)] + 14NaOH \rightarrow 5Na_4[Fe(CN)_6] + Fe(OH)^{2+}\ 6NaNO_3 + 6H_2O$

According to another view NO^+ group present in nitroprusside gets oxidised to NO_2 and thus a nitro complex is obtained.

$$Na_2[Fe(CN)_5(NO)] + 2NaOH \longrightarrow Na_4[Fe(CN)_5(NO_2)] + H_2O$$

(vii) $[Fe(CN)_5(NO)]^{2-}$ *ion is having diamagnetic character. Its diamagnetic character confirms the fact that* NO is *present as* NO^+ *ion is this complex ion.*

Uses : It finds use as a reagent in qualitative analysis for the detection of sulphide ion, aldehydes and ketones.

Structure: Sodium nitroprusside, $Na_2[Fe(CN)_5(NO)]^{2-}$, is having $[Fe(CNX)_5(NO)]^{2-}$ complex ion. In this ion iron is divalent (Fe^{2+}) and its electronic configuration is as :

$$26Fe^{2+} : 1s^2, 2s^22p^6, 3s^23p^6\ d^6$$

When Fe(II) forms complex ion with CN^- and NO^+ ligands, the 3d–electrons are paired and two 3d–orbitals become empty. Thus Fe(II) undergoes d^2sp^3 hybridization. Out of six d^2sp^3 hybrid orbitals five hybrid orbitals are occupied by five CN^- ligands [four CN^- ligand on the base and one CN^- ligand above or below the plane]. The sixth hybrid orbital (below or above the plane) is occupied by NO^+ ligands. Thus we can say $[Fe^{2+}(CN^-)_5(NO^+)]^{2-}$ ion has octahedral structure with Fe^{2+} ion situated at the centre of the octahedron and the complex ion is diamagnetic in nature. In this complex ion Fe–N–O angle is 178° *i.e.,* these bonds are not exactly collinear (≠ 180°) but just to linear. The Fe – N and N – O bond distanes are 1.63Å and 1.13 Å respectively.

Nitrosyl bis (diethyl dithiocarbamate $[Fe(NO)(S_2CNEt_2)_2]$ is another important nitrosyl complex which is prepared by dissolving $FeSO_4$ and $NaNO_2$ in dil. sulphuric acid and treating the solution with sodium diethyl dithiocarbamate. The solution which is of dark colour is extracted with chloroform and then from this solution the complex is crystallised on evaporation.

(F) Miscellaneous nitrosyl compounds: The best examples of this type are:

Roussins' black salts, $[M^IFe_4(NO)_7S_3]$, where $M^I = Na^+, K^+, Rb^+, Cs^+, NH_4^+, T_1^+$, etc. These compounds have been reported to be prepared by the action of iron (II) salt solutions on nitrite and sulphite ions.

Nitric oxide complexes of various types are known for Ru and Os. For ruthenium, nitric oxide complexes are very numerous more than for any other element and are a most important feature of ruthenium chemistry. The typical complexes are:

$$K_2[Ru(NO)(OH)(NO_2)_4],\ K_2[Ru(NO)(Cl)_5],$$
$$K_2[Ru(NO)(CN)_5]\ [Ru(NO)(NH_3)_4(Cl)]Cl_2 \text{ and}$$
$$[Ru(NO)(S_2CNR_2)_2]\ (R = CH_3)$$

These complexes can be obtained in a variety of ways.

STRUCTURE OF NITROSYLS CO-ORDINATION COMPOUNDS

If we compare the electronic structure of nitric oxide with carbon monoxide, it is observed that NO has an additional electron in antibonding π M.O. which may be readily lost to form the nitrosonium ion NO^+. With the result, the strength of N – O bond in NO should be increased by the loss of the antibonding electron. Actually, it has been confirmed by the increase in the N – O stretching frequency from 1878 cm^{-1} in free nitric oxide to 2200 – 2300 cm^{-1} in nitrosonium salts.

The additional electron present in π^*z molecular orbital of neutral nitric oxide can be supplied to metal atom thus increasing its effective number by one unit and neutral nitric oxide is itself coverted into NO^+ ion. Then, this NO^+ ion is co–ordinated through nitrogen with the metal atom by donating its lone pair to the metal.

(i) *Cobalt atom may increase its effective atomic number from* 27 *to* 28 *by accepting an additional electron from a neutral molecule of nitric oxide*

$$Co + NO \rightarrow CO^- + NO^+$$

Cobalt anion may then combine with one NO^+ group and three carbon monoxide molecules to form stable compounds $Co(NO)(CO)_3$.

$$Co^- + NO^+ + 3CO \rightarrow Co(NO)(CO)_3$$

In this compound, the effective atomic number of cobalt is 27 + 1 + 2 + 6 = 36 which is that of stable krypton gas.

(ii) *Similarly, the formation of* $Fe(NO)_2(CO)_2$ *can be explained.*

In 1934 Sidgwick and Bailey reported that the NO group is attached to the metallic atom by a link of two shared electrons; both being contributed by the nitrogen atom of nitric oxide group. Besides this, nitrogen atom loses its third electron to the metal, thereby, increasing its effective number by three. The electronic structure of metallic nitrosyls may be depicted as below.

$$M^{+} \longleftarrow, (:N: : :O:^{+}) \text{ or } M^{2-} - N^{+} \equiv O^{+}$$

The virtual accumulation of charge on the central atom favours strong π–bond formation with the attached groups.

Thus, the majority of metal nitrosyls are best regarded as being formed by donation from the nitrosonium ion (NO^{+}) to the metal atom with the M – O back–bonding in a manner entirely analogous to that in the M – C bond in carbonyl. In simple words, NO must be considered as a three electron donor:

$$M + NO \rightarrow M^{-} + NO^{+} \rightarrow M^{2-} - N^{+} = O^{+}$$

The infra-red and structural data clearly support the nitric oxide as three electron donor. Also, the structural studies of nitrosyls confirm that M–N–O groups are linear.

Since oxygen is more electronegative than nitrogen, it is nitrogen atom of NO^{+} ion which coordinates to M^{-} ion to form the nitrosyl complexes. In other words it means that the coordination of NO^{+} ion to NT ion occurs through the lone pair of electrons residing in the hybrid orbital on N atom *i.e.*, $(sp_z)^2$ *N* lone pair.

In terms of molecular orbital theory, the hybrid orbital on N atom having a lone pair [*i.e.*, $(sp_z]^2$ N lone pair] overlaps suitable vacant hybrid orbital on M ion in sp^3 in tetrahedral or d^2sp^3 in octahedral case to form $ON^{+} \longleftarrow M^{-}$ a bond. Now the empty π_z^* or π_y^* molecular orbital will overlap with the filled d–orbitals (d_{xy}, d_{yz} and d_{zx} orbitals in octahedral case and $d_x{}^2 - y^2$ and $d_z{}^2$ orbitals in tetrahedral case) to form $M^{-} \longrightarrow NO^{+}$ π bond. This type of overlap transfers some charge from M^{-} ion to NO^{+} ion.

Although, the concept of bonding involving three electrons accounts for the structures of many nitrosyl complexes, it suffers from the general criticism that an unlikely negative charge gets accumulated on the metal atoms.

An alternative is the four electron arrangement in which the metal acts as both a donor and an acceptor. However, this concept needs further discussion.

The molecule of nitric oxide may be regarded as a resonance structure of the following forms:

$$\dot{.N} \overset{\pi}{-} \underset{..}{O}: \quad \overset{+}{\underset{..}{\ddot{N}}} - \overset{-}{\underset{..}{\ddot{O}}}: \quad \overset{-}{\underset{..}{\ddot{N}}} \overset{\pi}{-} \overset{+}{\underset{.}{O}}:$$

On the basis of the resonance structures of nitric oxide, the metallic nitrosyls may be formulated as follows:

$$\underset{A}{R-\ddot{N}\overset{\pi}{-}\dot{\underset{\cdot}{O}}:}\quad \underset{B}{\overset{\pi}{R}-\overset{+}{N}\overset{\pi}{-}\dot{\underset{\cdot}{O}}:}\quad \underset{C}{\bar{R}-\overset{+}{N}\overset{\pi}{-}\dot{\underset{\cdot}{O}}:}\quad \underset{D}{\overset{-\pi}{R}-\overset{+}{\dot{N}}+\dot{\underset{\cdot}{O}}^{\pi}}$$

In the above structures, R is one atom of the metal.

(i) Iıı configuration B, one electron from nitrogen atom gets transferred to some orbital of the metal other than two orbitals which are engaged in bonding the nitrogen atom to the metal.

(ii) In configuration C, the nitrogen donates a lone pair of electrons to the metal atom and thus a co–ionic bond is established. This configuration is paramagnetic due to an unpaired electron on nitrogen atom.

Effective atomic number (EAN) rule as applied to metallic nitrosyls: Metallic nitrosyls also obey the EAN rule as demonstrated below for certain nitrosyls. In these cases NO is assumed to be a unipositive ion, NO^+ and hence acts ass a two–electron donor. Metal atoms are, therefore, in negative oxidation state.

(i) $[Co^-(Co)_3(NO)]^\circ$

$$\begin{aligned} Co^- &= 27e^- + 1e^- &&= 28e^- \\ 3CO &= 3 \times 2e^- &&= 6e^- \\ NO^+ &= 1 \times 2e^- &&= 2e^- \end{aligned}$$

$[Co^-(CO)_3(NO^+)]$ $= 36e$ (Kr)

(ii) $[Fe^{2-}(CO)_2(NO^+)_2]^\circ$

$$\begin{aligned} Fe^{2-} &= 26e^- + 2e^- &&= 28e^- \\ 2CO &= 2 \times 2e^- &&= 4e^- \\ 2NO^+ &= 2 \times 2e^- &&= 4e^- \end{aligned}$$

$[Fe^{2-}(CO)_2(NO^+)_2] = 36e^-$ (Kr)

(iii) $[Fe^{2-}(No^+)_2(PR_3)_2]^\circ$

$$\begin{aligned} [Fe^{2-} &= 26e^- + 2e^- &&= 28e^- \\ 2NO^+ &= 2 \times 2e^- &&= 4e^- \\ 2PR_3 &= 2 \times 2e^- &&= 4e^- \end{aligned}$$

$[Fe^{2-}\ (NO^+)_2(PR_3)_2] = 36e^-$ (Kr)

(iv) $[Mn^{3-}(CO)(NO^+)_3]^\circ$

$$\begin{aligned} Mn^{3-} &= 25e^- + 3e^- = 28e^- \\ CO &= 1 \times 2e^- &&= 2e^- \\ 3NO^+ &= 3 \times 2e^- &&= 6e^- \end{aligned}$$

$[Mn^{3-}(CO)(NO^+)_3] = 36e^-$ (Kr)

CARBONYLS

Since the electrons forming OC → M bond are supplied solely by CO molecules, metal atom in carbonyls is said to be in zero oxidation state. In metal carbonyls CO molecules act as neutral ligands. Metal carbonyls vary considerably in their properties ranging from volatile non-polar to the non-volatile electrovalent carbonyls. For example, nickel forms volatile non-polar carbonyls whereas alkali and alkaline earth metals from nonvolatile electrovalent carbonyls. In carbonyls, a metal atom is directly linked to the carbon atom of a carbonyl group. The general formula of the carbonyls may be given as $(M_xCO)_y$, where M is a metal capable of forming carbonyl. Metal carbonyls may be regarded as parents of a number of related compounds such as metal nitrosyl carbonyl $M(NO)_y\ (CO)_x$ and metal carbonyl hydrides, $H_xM(CO)_y$.

CLASSIFICATION OF CARBONYLS

Heiber, one of the most important workers in the field of carbonyls, classified the metal carbonyls into two distinct groups;

(a) Mono-nuclear Carbonyls

These carbonyls have the general formula $M(CO)_x$ which contain only one atom of the metal per molecule of the carbonyl. Thus, their empirical and molecular formula are identical. Examples are $Ni(CO)_4$, $Fe(CO)_5$, $Cr(CO)_6$, etc. It is important to remark that mononuclear carbonyls are generally formed by metals which possess even atomic numbers. Some characteristics of such carbonyls are :

(i) *These mononuclear carbonyls are readily soluble in organic solvents such as light petroleum, benzene, ether, etc.*

(ii) *These can be vaporised without decomposition.*

(iii) *These are more volatile than the others.*

(b) Polynuclear Carbonyls

These have the general formula $M_x(CO)_y$ which contain more than one metal atom per molecule. However some authors call the carbonyls having two metal atoms as bridged carbonyls and those having mote, than two metal atoms as *polynuclear carbonyls.*

Polynuclear carbonyls may be *homonuclear* [*e.g.*, $Fe_3(CO)_{12}$] or *heteronuclear* [*e.g.*, $MnCo(CO)_9$, $MnRe(CO)_{10}$]. Some characteristics of such carbonyls are outlined below :

(i) *These are almost insoluble in organic solvents.*

(ii) *Many polynuclear carbonyls decompose at or below the melting point.*

Compounds of transition metals possessing CO include carbonyls, carbonyl hydrides and carbonyl halides which are given in Table 2.1.

PREPARATION OF CARBONYLS

(a) Direct Synthesis from Metals and Carbonmonoxide

In practice this method is limited to carbonyls of those metals which can be obtained in active form by low-temperature reduction processess. For example,

(i) *Nickel reacts with carbon monoxide at room temperature and normal pressure.*

$$Ni + 4CO \xrightarrow{40°C} Ni(CO)_4$$

Table 3 : Metal Carbonyl, Carbonyl Hydrides and Carbonyl Halides

Group	*Elements*	*Carbonyl*	*Carbonyl hydrides*	*Carbonyl halides*
V	V	$V(CO)_6$		
VI	Cr, Mo, W	$M(CO)_6$	$Cr(CO)_5H_2$	
VII	Mn	$Mn_2(CO)_{10}$	$Mn(CO)_5H$	$Mn(CO)_5\ I$
				$Mn_2(CO)_8\ I_2$
				$M(CO)_5\ I$
	Tc, Re	$M_2(CO)_{10}$	$Re\ (CO)_5H$	$M(CO)_8\ I_2$
VIII		$Re_3(CO)_{12}$	$Fe(CO)_4\ H_2$	$Fe(CO)_2\ X_2$
	Fe	$Fe(CO)_5$	$Fe_3(CO)_{11}H_2$	$Fe(CO)_4X_2$
		$Fe_2(CO)_9$	$Fe(CO)_{13}\ H_3$	$Fe(CO)_4\ X_2$
		$Fe_3(CO)_2$	$Fe(CO)_5X_2$	
				$Fe_3(CO)_9\ X_6$
	CO	$Co_2(CO)_8$	$Co(CO)_4H$	$Co(CO)X_2$
		$Co_4(CO)_{12}$		
		$Co_6(CO)_{16}$	$Ni_2(CO)_6H_2$	
	Ni	$Ni(CO)_4$		

(ii) *When carbon monoxide is passed over reduced iron at* 780–220° *and pressure of* 50 *to* 200 *atm. pressure,* $Fe(CO)_3$ is *is formed.*

$$Fe + 5CO \rightleftarrows Fe(CO)_5$$

Rhenium, osmium and .indium carbonyls could not be prepared by direct reactions. Mittash in 1928 showed the reactivity of Fe, CO and Ni towards carbon monoxide in the decreasing order of Ni > Fe > Co. Generally, the direct combination of a metal with carbon monoxide is often catalysed by the presence of little sulphur or sulphur compound. For example, the combination of iron with CO at 200°C and 200 atomsphere pressure is very slow in the absence of sulphur. However, in the presence of H_2S the formation of $Fe(CO)_5$ becomes rapid even at 25 atmospheres. In the presence of sulphides, molybdenum and tungsten form carbonyls rapidly.

(b) Indirect Synthesis Involving the Grignard Reagent

Job prepared chromium hexacarbonyl by the action of carbon monoxide on a mixture of Grignard reagent and anhydrous chromium chloride in ether solution. Similar reactions in the presence of molydenum (V) and tungsten (V) chlorides gave the corresponding hexacarbonyls.

According to Heiber, fee primary reaction is as follows :

$$C_6H_5MgBr + CrCl_3 + CO \longrightarrow \underset{\text{Unstable}}{Cr(CO)_2(C_6H_5)_4} + MgBrCl + MgBr_2$$

The unstable intermediate .compound is decomposed with acid to yield the hexacarbonyl.

$$3Cr(CO)_2(C_6H_5)_4 + 6H^+ \longrightarrow Cr(CO)_6 + 2Cr^{3+} + 12C_6HI + 3H_2$$

These reactions are characterised by low yield which can be improved by using high carbon monoxide pressure. For example, if we use carbon monoxide at pressures of 35 to 70 atom, a 60% yield of the chromium hexacarbonyl is formed.

(c) Indirect Synthesis Involving Metal Compounds

Metal carbonyls can be prepared by the reaction of carbon monoxide with certain metal compounds. For example,

(i) *Nickel tetracarbonyl may be prepared by passing carbon monoxide into a suspension of nickel cyanide, sulphide or mercaptide suspended in sodium hydroxide solution or in general by treating nickel salts of inorganic or organic thioacids with carbon monoxide. Manchot and Gall suggested that this reaction proceeds by two steps :*

$$2NiX_4 + 2nCO \longrightarrow 2Ni(CO)_nX + X_2$$

$$2Ni(CO)_nX + (4 - 2n)\ CO \longrightarrow Ni\ (CO)_4 + NiX_2$$

(ii) *Ruthenium pentacarbonyl may be prepared by the action of carbon monoxide on* RuI_3 *in the presence of an iodine acceptor. First of all* $Ru(CO)_4I_2$ *is formed under ordinary pressure and at high pressure this compound is converted in to the pentacarbonyl.*

This may represent the following equilibrium :

$$RuI_3 \underset{}{\overset{[CO]}{\rightleftharpoons}} Ru(CO)_4I_2 \underset{}{\overset{[CO]}{\rightleftharpoons}} Ru(CO)_5$$

Similarly, $[lr(CO)_4]_7$ may be prepared.

(d) Synthesis by Carbonylating the Metallic Salts with CO in the Presence of Reducing Agent

When salts like RuI_3, $CrCl_3$ VCl_3, CoS, $CoCO_3$, CoI_2 etc, are made to treat with CO (carbonylation) in presence of a suitable reducing agent like Mg, Ag, Cu, Na, H_2, LiAllit etc, metallic carbonyls are obtained. For example :

$$CrCl_3 + CO + LiAlH_4 \xrightarrow{115°.\ 70\ atm.\ pressere} Cr(CO)_6 + LiCl + AlCl_3$$

$$2RuI_3 + 10CO + 6Ag \xrightarrow{175°C.\ 250\ atm.\ pressere} 2Ru(CO)_5 + 6AgI$$

$$\underset{\text{(in diethyl ether)}}{2MnI_2 + 10CO + 2Mg} \xrightarrow{25C°.\ 210\ atm.\ pressere} 2Mn_2(CO)_{10} + 2MgI_2$$

$$2CoS + 8CO + 4Cu \xrightarrow{200°C,\ 200\ atm\ pressure} Co_2(CO)_8 + 2Cu_2S$$

$$2FeS + 10CO + 2Cu \xrightarrow{200°C,\ 200\ atm\ pressure} 2Fe(CO)_5 + Cu_2S$$

$$2CoI_2 + 8CO + 4Cu \xrightarrow{200°C,\ 200\ atm\ pressure} Co_2(CO)_8 + 4CuI$$

$$FeI_2 + 8CO + 2Cu\ Fe(CO)_5 + Cu_2I_2$$

$$2CoCO_3 + 8CO + 2H_2 \xrightarrow{120°C, - 200°C,\ 250 - 300\ atm.\ pressure} Co_2(CO)_8 + 2CO_2 + 2H_2O$$

$$\underset{\text{(in pyridine)}}{2Cr(acac)_3} + 12CO + 3Mg$$

$$\xrightarrow{160°C,\ 300\ atm\ pressure} 3Mg(acac)_2 + 2Cr(CO)_6$$

$$MoCl_5 + 6CO + 5Na \xrightarrow{(in\ diglyme)} Mo(CO)_6 + 5NaCl$$

Sometimes CO is known to act both as a carbonylating and reducing agent. For example in the preparation of $Os(CO)_5$ and $Re_2(CO)_{10}$ from OsO_4 and Re_2O_7 respectively, CO acts both as a carbonylating and reducing agent.

$$OsO_4 + 5CO \xrightarrow{250°C,\ 350\ atm\ pressure} Os(CO)_5 + 2O_2$$

$$2Re_2O_7 + 10CO \longrightarrow Re_2(CO)_{10} + 7O_2$$

$V(CO)_6$ is obtained by the method represented by the following equation :

$$VC1_3 + 6CO + Na \xrightarrow[Acidification\ by\ H_3PO_4]{110°C,\ 150\ atm\ pressure} V(CO)_6 + 3NaCl$$

Here diglyme = diethyleneglycol dimethyl ether.

(e) Synthesis from Other Carbonyls

When iron pentacarbonyl is exposed to ultraviolet light, it loses carbon monoxide and forms $Fe_2(CO)_9$. This compound undergoes thermal decomposition to yield iron pentacarbonyl and trimeric tetracarbonyl.

$$2Fe(CO)_5 \xrightarrow{ultraviolet} Fe_2(CO)_9 + CO$$

$$2Fe_2(CO)_9 \xrightarrow{heat} Fe(CO)_5 + [Fe(CO)_4]_3 + CO$$

When osmium pentacarbonyl is heated, $Os_2(CO)_9$ is formed.

$$2Os(CO)_5 \xrightarrow{Heat} Os_2(CO)_9 + CO$$

Similarly, $Ru_2(CO)_9$ is obtained from $Ru(CO)_5$ when it is exposed to light.

(f) Synthesis from Carbonyl Hydrides

When iron carbonyl hydride is oxidised by MnO_2 or HaO_2, $[Fe(CO)_4]_3$ is formed. Dimeric tetracarbonyl of cobalt may be obtained by heating cobalt carbonyl hydride.

(g) By Treatment of Oxide of Metals with Carbonmonoxide Under Pressure

Carbonyls of osmium and rhenium are prepared by the reaction of carbon monoxide with their oxides under pressure.

$$OsO_4 + 9CO \xrightarrow[\text{50 Atm pressure}]{100°C} Os(CO)_5 + 4CO_2 \; Re_2(CO)_{10} + 7CO_2$$

(h) Preparation of $Mo(CO)_6$ and $W(CO)_6$ from $Fe(CO)_5$

Since CO groups present in $Fe(CO)_5$ are labile, they can be replaced by Cl^- ions by treating $Fe(CO)_5$ with MoClg or WCl_6 and thus $Mo(CO)_6$ and $W(CO)_6$ can be obtained. Thus :

$$MoCl_6 + 3Fe(CO)_5 \xrightarrow{100°C.\ ether} Mo(CO)_6 + 3FeCl_2 + 9CO$$

$$WCl_6 + 3Fe(CO)_5 \xrightarrow{100°C.\ ether} W(CO)_6 + 3FeCl_2 + 9CO$$

(i) Preparation of $Fe_2(CO)_9$ and $Os_2(CO)_9$ from $Fe(CO)_5$ and $Os(CO)_5$

When cooled solution of $Fe(CO)_5$ and $Os(CO)_5$ in glacial CH_3COOH is irradiated with ultraviolet light, $Fe_2(CO)_9$ and Os_2 $(CO)_9$ are obtained respectively.

$$2Fe(CO)_5 \xrightarrow{\text{ultra-violet light}} Fe_2(CO)_9 + CO$$

$$2Os(CO)_5 \xrightarrow{\text{ultra-violet light}} Os_2(CO)_9 + CO$$

Table 4 : Colour and melting points of some carbonyls

Carbonyl	*Colour and shape*	*Melting point. (°C)*
$V(CO)_6$	Black crystals	Decomposes at 70°C Sublime in vacuum
$Cr(CO)_6$	Colourless crystals	Sublime in vacuum
$Mo(CO)_6$	Colourless crystals	Sublime in vacuum
$W(CO)_6$	Colourless crystals	Sublime in vacuum
$Mn_2(CO)_{10}$	Golden crystals	154° – 155°
$Re_2(CO)_{10}$	Colourless crystals	Sublime at 140°and decompose at 177°C
$Fe(CO)_5$	Yellow liquid	B. P. 103°C
$Fe_2(CO)_9$	Bronze mica-lie plates	Decomposes at 100°C
$Fe_3(CO)_{12}$	Dark green crystals	Decomposes at ~140°C
$Co_2(CO)_6$	Orange crystals	51°C
$Ni(CO)_4$	Colourless liquid	B. P. 43°C

PROPERTIES OF CARBONYLS

Properties

(i) The metal carbonyls are crystalline solids, except for nickel carbonyl and the pentacarbonyls of iron, ruthenium and osmium which are liquids,

(ii) Many are coloured; for example, crystals of cobalt carbonyl are orange and Iron pentacarbonyl is a yellow oil, and nickel carbonyl is-colourless.

(ii) Their covalent nature renders them insoluble in water. Most of them are soluble in solvents like carbon tetrachloride.

(iv) Excepting $V(CO)_6$, all the carbonyls are diamagnetic. $V(CO)_6$ is paramagnetic and its paramagnetic property corresponds to the presence of one unpaired electron. The metals in carbonyls are in zero oxidation state.

Chemical

(i) *Substitution reactions : Some or all CO groups present in carbonyls can be replaced by monodentate ligands such as alkyl or aryl isocyanide* (CNR), PR_3, PCl_3, *py*, CH_3OH *etc. For example:*

$$Ni(CO)_4 + 4CNR \longrightarrow Ni(CNR)_4 + 4CO$$

$$Ni(CO)_4 + 4PC1_3 \longrightarrow Ni(PCl_3)_4 + 4CO$$

$$Fe(CO)_5 + 2CNR \longrightarrow Fe(CO)_3(CNR)_2 + 2CO$$

$$Mn_2(CO)_{10} + PR_3 \longrightarrow 2Mn(CO)_4(PR_3) + 2CO$$

$$2Fe_3(CO)_{12} + 3py \longrightarrow Fe_3CO)_9 + 3Fe(CO)_5$$

$$2Fe_3(CO)_{12} + 3CH_3OH \longrightarrow Fe_3(CO)_9(CH_3OH)_3 + 3Fe(CO)_5$$

$$Cr(CO)_6 \xrightarrow{+\,py} \underset{\text{yellow brown solid}}{Cr(CO)_4(py)_2}$$

$$\xrightarrow{+\,py} \underset{\text{yellow brown solid}}{Cr(CO)_4(py)_2}$$

$$\xrightarrow{+\,py} \underset{\text{bright red solid}}{Cr(CO)_2(py)_3}$$

Bidentate ligands like diars, o-phen, NO_2 etc are known to replace two or more CO groups, at a time. For example.

$$Mo(CO)_6 + \textit{diars} \longrightarrow Mo(CO)_4\ (\textit{diars}) + 2CO$$

$$Fe(CO)_5 + \textit{diars} \longrightarrow Fe(CO)_3\ \textit{diars} + 2CO$$

$$Ni(CO)_4 + \textit{o–phen} \longrightarrow Ni(CO)_2\ (\textit{o–phen}) + 2CO$$

$$Ni(CO)_4 + \textit{diars} \longrightarrow Ni(CO)_2\ (\textit{diars}) + 2CO$$

$$Cr(CO)_6 + 2\textit{diars} \longrightarrow Cr(CO)_2\ (\textit{diars})_2 + 4CO$$

$$Ni(CO)_4 + 2NO_2 \longrightarrow Ni(NO_2)_2 + 4CO.$$

(ii) *Action of* NaOH *or Na metal : Formation of carbonylate anion (Reduction). Aqueous or alcoholic solution of* NaOH *reacts with* $Fe(CO)_5$ *to form carbonylate anion,* $[HFe(CO)_4]^-$ $Fe(CO)_5$

$$(Fe = 0) + 3NaOH \longrightarrow Na^+[H^+Fe^{2-}CO)_4]^- + Na_2CO_3 + H_2O$$

H-atom in $[H^+Fe^{2-} + CO)_4$ ion is acidic which implies that Fe-atom in this ion is in – 2 oxidation sate.

Na-metal in liquid NH_3 is able to convert $Fe_2(CO)_9$, $Co_2(CO)_6$, $Fe_3(CO)_{12}$, $Cr(CO)_6$, $Mn_2(CO)_{10}$ etc. into carbonylate anions and in this conversion these carbonyls are reduced.

$$\underset{(Fe=0)}{Fe_2(CO)_9} + 4Na \longrightarrow 2Na_2^+\ [Fe^{2-}\ (C0)_4]^{2-} + CO$$

$$\underset{(Co=0)}{Co_2(CO)_8} + 2Na \longrightarrow 2Na^+[Co^-(CO)_4]^-$$

$$\underset{(Fe=0)}{Fe_3(CO)_{12}} + 3Na \longrightarrow 3Na^+\ [Fe^-(CO)_4]^-$$

$$\underset{(Cr=0)}{Cr(CO)_6} + 2Na \longrightarrow Na_2^+\ [Cr^{2-}(CO)_5]^{2-} + CO$$

$$\underset{(Mn=0)}{Mn_2(CO)_{10}} + 2Na \longrightarrow 2Na^+[Mn^-(CO)_5]^-$$

(iii) ***Action of halogens***: *Most of the carbonyls react with halogens to yield carbonyl halides. For example :*

$$Fe(CO)_5 + X_2 \longrightarrow Fe(CO)_4X_2 + CO$$

$$Mo(CO)_6 + Cl_2 \longrightarrow Mo(CO)_4Cl_2 + 2CO$$

$$Mn_2(CO)_{10} + X_2\ (X = Br, I) \longrightarrow 2Mn(CO)_5X$$

Both $Co_2(CO)_8$ and $Ni(CO)_4$ are decomposed into metallic halogens and CO when treated with halogens.

$$Co_2(CO)_8 + 2X_2 \longrightarrow 2CoX_2 + 8CO$$

$$Ni(CO)_4 + Br_2 \longrightarrow NiBr_2 + 4CO$$

(iv) ***Action of NO***: *Many carbonyls react with nitric oxide (NO) to form metal carbonyl nitrosyls. For example* :

$$Fe(CO)_5 + 2NO \xrightarrow{95°C} Fe(CO)_2(NO)_2 + 3CO$$

$$3Fe_3(CO)_9 + 4NO \longrightarrow 2Fe(CO)_2(NO)_2 + Fe(CO)_5 + Fe_3(CO)_{12} + 6\ CO$$

$$Fe_3(CO)_{12} + 6NO \xrightarrow{85°C} 3Fe(CO)_2(NO)_2 + 6CO$$

$$Co_2(CO)_8 + 2NO \xrightarrow{40°C} 2Co(CO)_3(NO) + 2CO$$

Moist NO yields a blue–coloured compound, Ni(NO)(OH) with $Ni(CO)_4$ while dry NO gives a blue solution of the composition, $Ni(NO)(NO_2)$

$$2Ni(CO)_4 + 2NO + 2H_2O \longrightarrow 2Ni(NO)(OH) + 8CO + H_2$$

$$Ni(CO)_4 + 4NO \longrightarrow Ni(NO)(NO_2) + 4CO + N_2O$$

(v) ***Action of HI***: *Formation of carbonyl hydrides (Reduction), When* M n(C O)$_{10}$ *and* $Co_2(CO)_8$ *react with* H_2 *they get reduced to carbonyl hydrides,* $Mn(CO)_5H$ *and* $Co(CO)_4H$ *respectively.*

$$Mn_2(CO)_{10} + H_2 \xrightarrow{200°C.\ 200\ atm.\ pressure} 2[Mn^-(CO)_5H^+]^o$$

$$Co_2(CO)_8 + H_2 \xrightarrow{165°C.\ 200\ atm.\ pressure} 2[Co^-(CO)_4H^+]$$

(vi) **Action *of heat***: *Different carbonyls yield different products when heated. For example* :

$$Fe(CO)_5 \xrightarrow{250°C} Fe + 5CO$$

$$\underset{\text{(in toluene)}}{3Fe_2(CO)_9} \xrightarrow{70°.\ Cool} 3Fe(CO)_5 + Fe_3(CO)_{12}$$

$$Fe_3(CO)_{12} \xrightarrow{140°.C} 3Fe + 12CO$$

$$2Co_2(CO)_8 \xrightarrow{50°\text{(inert atmosphere)}} Co_4(CO)_{12} + 4CO$$

$$Ni(CO)_4 \xrightarrow{180°C} Ni + 4CO$$

$$3Fe_2(CO)_9 \xrightarrow{50°C} 2Fe_3(CO)_{12} + 3CO$$

DIFFERENT METAL CARBONYLS

1. Carbonyls of Six B Metals

These form carbonyls of one type only $M(CO)_6$ where M = Cr, Mo or W. But chromium also forms $Cr(CO)_5$.

(A) Chromium Hexacarbonyl, $Cr(CO)_6$

Preparation

(i) It is prepared by Job's method by passing carbon monoxide at 50 atm. pressure and at room temperature into a suspension of chromic chloride in ether, which has been treated with phenyl magnesium bromide at – 70°C.

(ii) Chromium hexacarbonyl can be prepared by treating a solution of a chromic salt dissolved in ether with $Al(C_2H_5)$ and carbon monoxide at a high temperature and pressure.

(iii) It may also be prepared by treating a chromium acetylacetone complex salt or a chromous or a chromic salt of an organic acid in pyridine solution with carbon monoxide at 100 – 300 atm pressure and 80 – 170°C in the presence of powdered zinc or magnesium.

The yield of this method is about 85%.

(iv) It may also be prepared by carbonylating $CrCl_3$ with CO in the presence of a reducing agent like lithium aluminium hydride.

$$CrCl_3 + CO + LiAlH_4 \xrightarrow{175°C.\ 70\text{ atm pressure}} Cr(CO)_6 + LiCl + AlCl_3$$

Properties

(i) *Chromium hexacarbonyl exists in colourless rhombic crystals which sublime without decomposition and dissolve in ether, chloroform, carbon tetrachloride and benzene.*

(ii) *It is not attacked by air, bromine, cold aqueous alkalies, dilute acids, conc. HCl and conc.* H_2SO_4. *It is, however, decomposed by chlorine or by conc, nitric acid.*

(iii) ***Decomposition:*** *It gets decomposed by* F_2 *at* – 75°C *to form* CrF_6.

(iv) ***Action of Na-metal or NaBH$_4$ in liquid NH$_3$:*** *Formation of carbonylate union (Reduction).*

When $Cr(CO)_6$ *is treated with Na metal or* $NaBH_4$ *in liq.* NH_3. *carbonylate union is formed.*

In these reactions the carbonyl is reduced.

$$Cr(CO)_6\ (Cr = 0) + 2Na \xrightarrow{\text{liq. } NH_3} Na_2^+[Cr^{2-}(CO)_5]^{2-} + CO$$

$$Cr(CO)_6(Cr = 0) \xrightarrow{NaBH_4/\text{liq. } NH_3} Na_2^+\ [Cr^{2-}(CO)_{10}]^{2-} + 2CO$$

(v) ***Substitution reactions:*** *Some CO groups present in* $Cr(CO)_6$ *can be replaced by pyridine (py) to get a number of products.*

$$Cr(CO)_6 \xrightarrow{+py} \underset{\text{yellow borown solid}}{Cr(CO)_4(py)_2}$$

$$\xrightarrow{+py} \underset{\text{yellow red solid}}{Cr_2(CO)_7(py)_5}$$

$$\xrightarrow{+py} \underset{\text{bright red solid}}{Cr(CO)_3(py)_3}$$

(vi) *When chromium hexacarbonyl is treated with potassium hydroxide in a methyl alcohol solution in a sealed tube at* 100°C *it yields a red solution due to the following reaction.*

$$Cr(CO)_6 + KOH + 5H_2O \longrightarrow \underset{\text{Red solution}}{K[Cr(CO)_3(H_2O)_2OH]} + 3HCOOH$$

(B) Chromium Pentacarbonyl, *Cr(Cp)$_5$*

The evidence for this five–coordinate species has been obtained by ultraviolet irradiation of the hexacarbonyl solutions; new bands are found in the C – O stretching region which suggests that the species may have a square based pyramidal structure [rather than the trigonal bipyramidal shape of $Fe(CO)_5$]. The electronic spectrum of the solution has been measured.

(C) Molybdenum Hexacarbonyl and Tungsten Hexacarbonyl

Preparation

(i) *Both these carbonyls may be prepared by Job's method which involves the reaction of either* $MoCl_6$ *or* WCl_6 *with carbon monoxide in the presence of phenyl magnesium bromide.*

(ii) *Both may also be prepared by the action of carbon monoxide at 225° and 200 atm. pressure on metallic molybdenum or tungsten reduced in the presence of copper or iron.*

(iii) $Mo(CO)_6$ *and* $W(CO)_6$ *may be obtained by the action of carbon monoxide on the metal in the presence of sulphide of copper or iron.*

(iv) $Mo(CO)_6$ *and* $W(CO)_6$ *can be prepared by treating the compounds* K_3MoCl_9, K_3WCl_9 *with carbon monoxide.*

Properties

(i) *They are colourless crystals.* $Mo(C\ O)_6$ *sublimes at 40° and boils at* 156.4° *whereas* $W(CO)_6$ *sublimes at 50° and boils at* 17 5°C.

(ii) *They are stable in air and dissolve in organic solvents like ether, chloroform, carbon tetrachloride and benzene.*

(iii) $Mo(CO)_6$ *and* $W(CO)_6$ *do not react with air, cold aqueous alkalies, acids except cone, nitric acid or with thiols or nitric oxide.*

(iv) *Bromine and chlorine can decompose* $Mo(CO)_6$ *and* $W(CO)_6$.

(v) *With pyridine, phenanthroline and ethylene diamine, the carbon monoxide group in* $Mo(CO)_6$ *and* $W(CO)_6$ *is replaced.*

$$M(CO)_6 \xrightarrow[140°C]{\text{Pyridine}} M(CO)_5Pyr_2 \longrightarrow M_2(CO)_7Pyr_5 \downarrow M(CO)_3Pyr_3$$

where M = Mo and W

(vi) *Both react with sodium in liquid ammonia corresponding to that of chromium hexacarbonyl.*

$$W(CO)_6 + 2Na \longrightarrow Na_2\,[W(CO)_5] + CO$$

$$Mo(CO)_6 + 2Na \longrightarrow Na_2\,[Mo(CO)_5] + CO$$

(vii) *Both have octahedral configuration.*

2. Carbonyls of VII Group

These form volatile carbonyls of the formula $M_2(CO)_{10}$ where M = Mn, To and Re

Manganese Carbonyl, $Mn_2\,(CO)_{10}$

(i) *This is prepared by treating manganese iodide and magnesium with carbon monoxide in ether under high pressure. In this reaction, magnesium acts as a reducing agent.*

$$2MnI_2 + 10\ CO + 2Mg\ \text{(in diethyl ether)}$$

$$\xrightarrow{25°C,\ 210\ \text{atm.pressure}} Mn_2(CO)_{10} + 2MgI_2$$

(ii) *By carbonylating* $MnCl_2$ *with CO in presence of* $(C_6H_5)_2CONa$

$$2MnCl_2 + 10CO + 4(C_6H_5)_2CONa \xrightarrow[\text{pressure}]{165°C,\ 140\ \text{atm.}}$$

$$Mn_2(CO)_{10} + 4(C_6H_5)_2CO + 4NaCl$$

Properties

(i) *Manganese carbonyl forms volatile golden yellow crystalline solid which melts at 155° in a sealed tube. It is soluble in organic solvents. It is slowly oxidised in air, especially in solution.*

(ii) ***Action of halogens***: $Mn_2(CO)_{10}$ *reacts with halogens to form carbonyl halides,*

$$Mn(CO)_5X.\ Mn_2(CO)_{10} + X_2 (X = Br, I) \longrightarrow 2Mn(CO)_5X$$

(iii) ***Action of Na–metal*** : *Na-metal in liquid* NH_3 *converts* $Mn_2(CO)_{10}$ *into carbonylate onion. In this reaction the oxidation state of Mn decreases from zero to –1.*

$$Mn_2(CO)_{10}(Mn = 0) + 2Na \longrightarrow 2Na^+\ [Mn^-(CO)_5]^-$$

(iv) *Action of* H_2: $Mn_2(CO)_{10}$ *gives carbonyl hydride,* $Mn(CO)_5H$. *In the formation of this compound the oxidation state of Mn decreases from zero to – 1.*

$$Mn_2(CO)_{10}(Mn{=}0) + H_2 \xrightarrow{200°C,\ 200\ \text{atm pressure}} 2[Mn^-(CO)_5H^+]°$$

(v) ***Substitution reaction***: $Mn_2(CO)_{10}$ *reacts with* PR_3 *to form* $Mn(CO)_4(PR_3)$.

$$Mn_2(CO)_{10} + PR_3 \longrightarrow 2Mn(CO)_4(PR_3) + 2CO$$

(vi) ***Diamagnetic nature***: $Mn_2(CO)_{10}$ *is a diamagnetic substance. Diamagnetic character confirms the facts that all the electrons in* $Mn_2(CO)_{10}$ *are paired and Mn–Mn bond is also present in it.*

Technetlum Carbonyl, $Tc_2(CO)_{10}$

In the year 1961 Higgins and co–workers reported the formation of the dimeric carbonyl of technetium as $Tc_2(CO)_{10}$. This carbonyl is similar to manganese and rhenium carbonyls. Its various iodine derivatives have also been prepared.

Rhenium Carbonyl, $Re_2(CO)_{10}$

The best method for preparing rhenium carbonyl is by the reaction of rhenium heptoxide with carbon monoxide at 250°C and 200 atm.

$$Re_2O_7 + 17CO \longrightarrow Re_2(CO)_{10} + 7CO_2$$

It is also prepared less easily by the reaction of potassium perrhenate and rhenium heptasulphide Re_2S_7 with carbon monoxide under pressure. Rhenium carbonyl forms colourless monoclinic crystals. It sublimes without decomposition at 140° but decomposes above the melting point 177°.

It is attacked by halogens to produce rhenium carbonyl halides $Re(CO)_5X$. It is not attacked by alkalies or cold mineral acids. However, it is attacked by hot cone, nitric acid or sulphuric acid. With sodium, it reacts to give $Na^+ [Re(CO)_5]^-$. It reacts with amines to form $Re(CO)_3 Pyr_2$; etc.

3. Carbonyls of VIII Group Metals

(A) Carbonyls of Iron

Three carbonyls of iron are known. These are :

(a) Iron Pentacarbonyl, Fe $(CO)_5$

Preparation

(i) *It can be prepared by the action of carbon monoxide on iron powder at* 200°C *and* 200 *atm. pressure.*

$$Fe + 5CO \longrightarrow Fe(CO)_5$$

(ii) *Recently, it has been prepared by the action of carbon monoxide on ferrous iodide in the presence of copper which acts as a halogen acceptor.*

$$FeI_2 + 4CO \xrightarrow[\text{200 atm. Pressure}]{200°C} Fe(CO)_4I_2$$

$$CO + Fe(CO)4I_2 + 2Cu \longrightarrow Fe(CO)_5 + Cu_2I_2$$

(iii) *It may also be prepared by the action of carbon monoxide an FeS at* 200°C *and* 200 *atm. pressure in the presence of copper.*

$$2FeS + 10CO + 2Cu \xrightarrow{\text{200°C, 200 atm. pressure}} 2Fe(CO)_5 + Cu_2S$$

Properties

(i) $Fe(CO)_3$ *is a yellow liquid which is soluble in methyl alcohol, ether, acetone and* C6H6. It *is insoluble in* H_2O.

(ii) **Decomposition:** *On thermal decomposition at* 250°C, *it yields pure Fe.*

$$2Fe(CO)_5 \xrightarrow{250°C} Fe + SCO$$

(iii) ***Action of ultra-violet light:*** *When cooled solution* of $Fe(CO)_5$ *in glacial* CH_3COOH *is irradiated with ultra violet light,* $Fe_2(CO)_9$ *is formed.*

$$2Fe(CO)_5 \xrightarrow{\text{ultra-violet light}} Fe_2(CO)_9 + CO$$

The above reaction is reversed in darkness.

(iv) ***Hydrolysis:*** $Fe(CO)_5$ *gets hydrolysed by* H_2O *and acids.*

$$Fe(CO)_5 + H_2SO_4 \longrightarrow FeSO_4 + SCO + H_2$$

(v) ***Action of alkalies:*** *"hen heated with aqueous or alcoholic solution of NaOH, carbonylate anion,* $[Fe^{2-}(CO)_4H^+]^-$ *is formed. The hydrogen atom in this anion is acidic which implies that Fe is in* – 2 *oxidation state.*

$$\underset{(Fe=0)}{Fe(CO)_5} + SNaOH \longrightarrow Na^+ \underset{(Fe=-0)}{[Fe^{2-}(CO)_4H+]^-} + Na_2CO_3 + H_2O$$

(vi) ***Action of NH_3:*** *With* NH_3 *it yields* $Fe(CO)_4H_2$ *and carbamic acid,* NH_2COOH

$$Fe(CO)_5 + H_2O + NH_3 \longrightarrow Fe(CO)_4H_2 + NH_2COOH$$

(vii) ***Reaction with halogens:*** *It reacts with halogens in nonaqueous solvents to form the stable tetra-carbonyl halides,* $Fe(CO)_4X_2$.

$$Fe(CO)_5 + X_2 \longrightarrow Fe(CO)_4X_2 + CO$$

The velocities of these reactions have been found to follow the order Cl < Br < I.

viii) *Reactions with ethylene diamine (en) and pyridine (py). When* $Fe(CO)_5$ *reacts with ethylene diamine (en), an addition product,* $Fe(CO)_5$ *(en) is formed. Pyridine (py) reacts with it to farm* $[Fe(py)_6]$ $[Fe_4(CO)_{13}]$.

$$Fe(CO)_5 + en \longrightarrow Fe(CO)_5(en)$$

$$5Fe(CO)_5 + 6py \longrightarrow [Fe(Py)_6]\,[Fe_4(CO)_{13}] + 12CO$$

(ix) ***Reaction with NO***: *Iron carbonyl nitrosyl,* $Fe(CO)_2(NO)_2$ *is formed when* $Fe(CO)_5$ *reacts with NO at* 95 °C.

$$Fe(CO)_5 + 2NO \longrightarrow Fe(CO)_2(NO)_2 + 3CO$$

(x) ***Action on metallic and non-metallic halides***: *The reactions occurring between* $Fe(CO)_5$ *and metallic or non-metallic halides can be divided into two categories*:

(a) When $Fe(CO)_5$ reacts with the halides like CCl_4, SO_2Cl_2, $SnCl_4$ $SbCl_5$ and $CuCl_2$, halides or carbonyl halides of Fe (II) get formed. Thus, the oxidation state of Fe increases from zero to + 2. For example:

$$Fe(CO)_5 + 2CCl_4 \longrightarrow Fe^{2+}Cl_2 + C_2Cl_6 + 5CO$$

$$Fe(CO)_5 + SO_2Cl_2 \longrightarrow Fe^{2+}Cl_2 + SO_2 + 5CO$$

$$Fe(CO)_5 + SnCl_4 \longrightarrow Fe^{2+}(CO)_4Cl_2 + SnCl_2 + CO$$

$$Fe(CO)_5 + SbCl_5 \longrightarrow Fe^{2+}(CO)_4Cl_2 + SbCl_3 + CO$$

$$Fe(CO)_5 + 2CuCl_2 \longrightarrow Fe^{2+}Cl_2 + 2CuCl + 5CO$$

(b) When $Fe(CO)_5$ reacts with WCl_6 or $MoCl_6$ all the CO groups present in $Fe(CO)_5$, being labile, get replaced by Cl–atoms. Thus, in this reaction, $Fe(CO)_5$ acts as a dehalogenating and carbonylating agent. Oxidation state of Fe gets increased from zero to + 2 (in $FeCl_2$).

$$MCl_6(M = W, Mo) + Fe(CO)_5 \xrightarrow{100°C,\ ether} M(CO)_6 + 3FeCl_2 + 9CO$$

(xi) ***Action of air:*** *When* $Fe(CO)_5$ *is burnt in air,* Fe_2O_3 *is formed.*

$$4Fe(CO)_5 + 3O_2 \xrightarrow{Burn} 2Fe_2O_3 + 20CO$$

(b) Iron Enneacarbonyl, $Fe_2(CO)_9$

When iron pentacarbony is dissolved in glacial acetic acid and is exposed to ultraviolet light for 6 hours, $Fe_2(CO)_9$ is formed which dissolves in acetic acid. On cooling with water, golden crystals of the enneacarbonyl are precipitated, and are filtered off.

$$2Fe(CO)_5 \rightleftarrows Fe_2(CO)_9 + CO$$

Properties

(i) $Fe_2(CO)_9$ *forms golden triclinic crystals. It is diamagnetic and non-volatile. It is insoluble in water but soluble in toluene and pyridine. When* $Fe_2(CO)_9$ *is heated to* 50°C, *it decomposes to form* $Fe_3(CO)_{12}$.

$$3Fe_2(CO)_9 \longrightarrow 3Fe(CO)_5 + Fe_3(CO)_{12}$$

(ii) ***Action of heat:*** *When heating is done at* 100°C, $Fe_2(CO)_9$ *decomposes to form iron, carbon monoxide and some* $Fe(CO)_5$ *and* $Fe_3(CO)]_{12}$.

$$4Fe_2(CO)_9 \longrightarrow Fe + Fe(CO)_5 + Fe_3(CO)_{12} + CO$$

This reaction is not quantitative. When $Fe_2(CO)_9$ is heated with nitric oxide at 75 – 100°C, it forms $Fe(CO)_2(NO)_2$ also with $Fe(CO)_5$ and $Fe_3(CO)_{12}$; $Fe_2(CO)_9$ undergoes reaction with o–phenanthroline in benzene at 80°C to form [Fe $Phen_3$] $[Fe(CO)_8]$

(iii ***Action of NO:*** *With NO it gives* $Fe(CO)_2(NO)_2$ *together with* $Fe(CO)_5$ *and* $Fe_3(CO)_2$.

$$3Fe_2(CO)_9 + 4NO \longrightarrow 2Fe(CO)_2(NO)_2 + Fe(CO)_5 + Fe_3(CO)_{12} + 6CO$$

(iv) ***Action of phen:*** *In acetone or benzene solution at* 80° $Fe_2(CO)_9$ *reacts with phen to form.* $[Fe(phen)_3]$ $[Fe_2(CO)_8]$.

In pyridine, with phen it gives $[Fe(phen)_2]$ $[Fe_4(CO)_3]$.

(v) ***Action of alkali metals:*** Carbonylate onion is obtained in its reaction with Na in NH_3

$$Fe_2(CO)_9(Fe = 0) + 4Na \xrightarrow{\text{liq. } NH_3} 2Na_2^+ [Fe^{2-}(CO)_4]^{2-} + CO$$

(c) Iron Dodecacarbonyl, $Fe_3(CO)_{12}$

Preparation

It can be prepared by heating $Fe_2(CO)_9$ dissolved in toluene at 70°C

$$3Fe_2(CO)_9 \longrightarrow 3Fe(CO)_5 + Fe_3(CO)_{12}.$$

On cooling rhe green crystals $Fe_3(CO)_{12}$ separate out.

$Fe_3(CO)_{12}$ may also be obtained by the oxidation of iron carbonyl hydride by hydrogen peroxide or MnO_2

$$3Fe(CO)_4H_2 + 3H_2O_2 \longrightarrow Fe_3(CO)_{12} + H_2O$$

$$3Fe(CO)_4H_2 + 3MnO_2 + 3H_2SO_4 \longrightarrow Fe_3(CO)_{12} + 3MnSO_4 + 6H_2O$$

Iron dodecacarbonyl can also be prepared by the action of acids on iron carbonyl hydride.

Properties

(i) $Fe_3(CO)]_{12}$ *forms deep green crystals which are soluble in organic solvents like toluene, alcohol, ether and pyridine.*

(ii) ***Action of heat***: *When it is heated to* 140°C $Fe_3(CO)_{12}$ *decomposes to give metallic iron and CO.*

$$Fe_3(CO)_{12} \xrightarrow{140°C} 3Fe + 12CO$$

(iii) ***Reaction with Na***: *Carbonylate anion is formed when* $Fe_3(CO)_{12}$ *reacts with Na metal in Liq.NH_3.*

$$Fe_3(CO)_{12} \xrightarrow{\text{liq. } NH_3} 3Na_2^+[Fe^{2-}(CO)_4]^{2-}$$

(iv) ***Substitution reactions***: *These reaction take place with pyridine (py) and methyl alcohol as shown below:*

$$2Fe_3(CO)_{12} + 3py \longrightarrow Fe_3(CO)_9(py)_3 + 3Fe(CO)_5$$

$$2Fe_3(CO)_{12} + 3CH_3OH \longrightarrow Fe_3(CO)_9(CH_3OH)_3 + 3Fe(CO)_5$$

(v) ***Action of NO***: *At 85° it reacts with NO to form iron dicarbonyl dinitrosyl,* $Fe(CO)_2(NO)_2$.

$$Fe_3(CO)_{12} + 6NO \longrightarrow 3Fe(CO)_2(NO)_2 + 6CO$$

Carbonyls of Cobalt

It forms two carbonyls,

Cobalt Octacarbonyl $CO_2(CO)_6$

Preparation

(i) *It is prepared by the action of carbon monoxide on the reduced metallic cobalt at 220°C and 250 atm*

$$2CO + 8CO \longrightarrow CO_2\ (CO)_6$$

(ii) *It is best prepared by the action of dry carbon monoxide at 200°C and 700 atm. pressure in the presence of metallic copper on certain binary compounds of cobalt, (e.g., CoS)*

$$2CoS + 8CO + 4Cu \longrightarrow Co_2(CO)_8 + 2Cu_2S$$

(iii) *When a solution of cobalt carbonyl hydride is treated by an acid, hydrogen is evolved and $CO_2\ (CO)_8$ remains.*

$$2Co(CO)_4H \longrightarrow CO_2(CO)_8 + H_2$$

(iv) *When cobalt carbonyl hydride is cooled to –79° in a closed vessel and is then allowed to warm up to about – 0°C it undergoes decomposition to yield $CO_2(CO)_8$. The vessel is recooled and the liberated hydrogen is pumped off.*

$$2Co(CO)_4H \longrightarrow Co_2(CO)_8 + H_2$$

Properties

(i) *Cobalt octacarbonyl forms orange transparent crystals, mp =* 51°C.

(ii) *It is insoluble in water but is soluble to some extent in alcohol, ether, carbon disulphide, etc.*

(iii) *Either pure or in solution it starts decomposing on standing or at temperatures about 51° to yield black crystals of* $[Co(CO)_3]_4$. *When the temperature is above 60°C, the black crystals decompose themselves to yield carbon monoxide and metallic cobalt.*

$$2Co_2(CO)_8 \xrightarrow{50°C} Co_4(CO)_{12} + 4CO$$

(iv) **Action of air:** *On exposure to air, dicobalt octacarbonyl is converted into deep violet basic carbonate of cobalt.*

(v) ***Action of H_2:*** *Formation of cobalt carbonyl hydride (Reduction). When* $Co_2(CO)_8$ *reacts with* H_2 *at* 165°C *and* 200 *atm. pressure, it gets reduced to cobalt carbonyl hydride,* $[Co^-(CO)_4H^+]^0$.

$$Co_2(CO)_8 + H_2 \xrightarrow{165°C.\ 200\ atm.\ presssure} 2[Co^-CO)_4H^+]^0$$

(vi) ***Action of Na-metal in llq.* NH_3:** *Reduction. When* $Co_2(CO)_8$ *reacts with Na–metal in liq. Na–metal below* – 75°C, *it gets reduced to carbonylate union,* $[Co^-(CO_4)]^-$.

$$Co_2(CO)_8 + 2Na \xrightarrow{liq.\ NH_3\ +} 2Na^+\ [Co–(CO)_4]^-$$

(vii) ***Disproportionate reactions:*** *There are two types of such reactions in case* $Co_2(CO)_8$.

(a) Strong bases having nitrogen or oxygen donor atoms cause disproportionation into Co(+2) and Co(– l). For example:

$$\underset{(Co\,=\,o)}{2Co_2(CO)_8} + 12NH_3 \longrightarrow \underset{(Co\,=\,+2)}{2[Co(NH_3)_6]}\ \underset{(Co\,=\,-1)}{[Co(CO)_4]_2} + 4CO$$

(b) With isocyanides, phosphines, arsines and stibines, however, the disproportionation reaction gives penta–coordinate cobalt (I) cation.

$$\underset{(Co\,=\,0)}{Co_2(CO)_8} + 5CNR \longrightarrow \underset{(Co\,=\,+1)}{[Co_2(CNR)_5]}\ \underset{(Co\,=\,-1)}{[Co_2(CO)_4]} + 4CO$$

(viii) *Action of NO.* $Co_2(CO)_8$ *reacts with NO at* 40°C *to form cobalt carbonyl nitrosyl,* $[Co^-(CO)_3(NO)^+]^0$. *Thus in this reaction the oxidation state of cobalt decreases from zero (in corbonyl) to* –1 *(in nitrosyl).*

$$\underset{(Co\,=\,0)}{Co_2(CO)_8]} + 2NO \longrightarrow \underset{(Co\,=\,-1)}{[Co^-(CO)_8(NO)^+]^0} + 2CO$$

(ix) ***Action of halogens:*** *The halogens* (X_2) *decompose* $Co_2(CO)_8$ *into* CoX_2 *and CO according to the following equation.*

$$Co_2(CO)_8 + 2X_2 \longrightarrow 2CoX_2 + 8CO$$

Cobalt Dodecacarbonyl, $Co_3(CO)_{12}$

(i) *It is prepared by warming cobalt octacarbonyl to* 60°C.

(ii) *It is also prepared by oxidizing cobalt carbonyl hydride below-* 20°C.

Properties

It forms black crystalline compound. It decomposes at 60°C. It is very unstable and is easily oxidised by air.

DODECACARBONYLTETRACOBALT (O), $[CO_4(CO)_{12}]$

Preparation

(i) *It is prepared by heating* $Co_2(CO)_8$ *at* 60°C.

(ii) *It may also be obtained by oxidizing cobalt carbonyl hydride below* –26 °C.

Properties

(i) *It is a black crystalline solid.*

(ii) *It is very unstable compound easily oxidized by air and can be recrystallized from hot benzene.*

(iii) *It is chemically inert and does not react with nitric oxide.*

Carbonyl of Nickel

Nickel Tetracarbonyl, Ni(CO)$_4$

It is prepared as follows :

(i) ***By direct synthesis:*** *Nickel tetrucarbonyl can be prepared by the action of carbon monoxide on reduced nickel at* 30–50°C

$$Ni + 4CO \longrightarrow Ni(CO)_4$$

The reduced nickel is prepared by the reduction of nickel oxalate or hydroxide by carbon monoxide at temp, below 300°C.

(ii) ***From nickel iodide:*** *When, nickel iodide is heated with carbon monoxide in the presence of a halogen acceptor, nickel carbonyl is formed.*

$$NiI_2 + 4CO \longrightarrow Ni(CO)_4 + I_2$$

(iii) ***From the salt of iron carbonyl halide and the nickel hexammine cation:*** *Nickel carbonyl is formed when carbon monoxide is passed over a salt of iron carbonyl halide and nickel hexammine cation.*

$$[Ni(NH_3)_6]\,[Fe(CO)_4H]_2 + 4CO \longrightarrow [Fe(CO)_4H]_2[Ni(CO)_4] + 6NH_3$$

$$4[Fe(CO)_4H]_2[Ni(CO)_4] \longrightarrow 3Fe(CO)_4H_2 + Fe_3(CO)_2 + 3Ni(CO)_4$$

(iv) ***From nickel salt:*** *Nickel carbonyl can be prepared by the action of carbon monoxide on an alkaline suspension of nickel sulphide or cyanide.*

$$NiS + 4CO \longrightarrow Ni(CO)_4 + S$$

$$Ni(CN)_2 + 4CO \longrightarrow Ni(CO)_4 + (CN)_2$$

$$S + 4OH^- + CO \longrightarrow S2^- + CO_3^{2} + 2H_2O$$

Properties

(i) *It is a colourless liquid having m.p.* = – 23°C, *b.p.* = 43°C.

(ii) *It has no solubility in water but disssolves readily in organic solvents.*

(iii) ***Action of heat:*** *It decomposes at* 180 – 200°C *to form nickel and carbon monoxide.*

$$Ni(CO)_4 \xrightarrow{180°C} Ni + 4CO$$

(iv) ***Action of halogens:*** *It reacts with gaseous chlorine to form nickel carbonyl halide. Similarly, it reacts with bromine. In organic solvents, halogens react to form nickel halide and carbon monoxide,*

$$Ni(CO)_4 + Cl_2 \longrightarrow NiCl_2 + 4CO$$

$$Ni(CO)_4 + Br_2 \longrightarrow NiBr_2 + 4CO$$

(v) ***Action of acids:*** *Aqueous halogen acids have no action. However, aqueous sulphuric acid reacts thus:*

$$Ni(CO)_4 + H_2SO_4 \longrightarrow NiSO_4 + H_2 + 4CO$$

Cone, sulphuric acid reacts with detonation.

(vi) ***Action of Ba(OH)$_2$:*** *It reacts with* $Ba(OH)_2$

$$Ni(CO)_4 + Ba(OH)_2 \longrightarrow H_2Ni(CO)_3 + BaCO_3$$

(vii) ***Action of nitric oxide:*** *With moist nitric oxide, it reacts to give a compound of intense blue colour Ni(NO)(OH).*

(viii) ***Action of sulphur and hydrogen sulphide:*** *Nickel carbonyl reacts with sulphur and with hydrogen sulphide.*

Carbonyls of Ruthenium

It forms three carbonyls.

(a) Ruthenium Pentacarbonyl, $Ru(CO)_6$

Preparation

(i) *It is prepared by the action of carbon monoxide on reduced ruthenium at 200°C and 200 atm. pressure.*

$$Ru + 5CO \longrightarrow Ru(CO)_5$$

(ii) *It is also prepared by the action of carbon monoxide on* RuI_3 *mixed with finely divided silver at* 170° *and* 450 *atm, pressure.*

Properties

(i) *It is a colourless soluble liquid having m.p,=* – 22'C.

(ii) *It has no solubility in water but is soluble in alcohol, benzene and chloroform.*

(iii) *It undergoes decomposition to give* $Ru_2 (CO)_9$ and $Ru_3 (CO)_{12}$.

(iv) *It reacts with halogens to yield Ru (CO) Br and carbon monoxide.*

(v) *It is photosensitive and yields ruthenium enneacarbonyl.*

(b) Ruthenium Enneacarbonyl, $Ru_3(CO)_9$: It is prepared by exposing pentacarbonyl to ultraviolet radiation. It may be prepared by heating a solution of pentacarbonyl in benzene at 50°C. It forms yellow monoclinic crystals. It is volatile. It is less stable towards heat. With iodine, it yields $Ru(CO)_2I_2$. With nitric oxide, it reacts to form $Ru(NO)_5$.

(c) $Ru_3(CO)_{12}$: It is prepared in small quantities along with $Ru_2(CO)_9$ when $Ru(CO)_5$ is heated at 50°C or by exposing $Ru(CO)_5$ to ultraviolet light. It is a green crystalline solid.

Rhodium Octacarbonyl $Rh_2(CO)_8$: It is prepared by the action of carbon monoxide on metallic rhodium under high pressure or by the action of carbon monoxide on rhodium tri–iodide at 100°C and high pressure in the presence of halogen acceptor, such as metallic copper. Rhodium octacarbonyl is an orange crystaline solid having m.p. = 76°C.

$Rh_4(CO)_{11}$: It is a black crystalline solid and is obtained by treating rhodium trichloride with carbon monoxide under pressure below' 100'C in the presence of copper which is a halogen acceptor.

$Rh_4(CO)_{11}$ is chemically inert. It decomposes at 150°C.

Carbonyls of Osmium

It forms two carbonyls;

(a) Osmium Pentacarbonyl, $Os(CO)_5$: It is a colourless liquid having m.p. = – 15°C. It is obtained.

(i) *By the action of carbon monoxide on* OsI_3 *at* 120°C, *and* 200 *atm pressure in the presence of copper.*

(ii) *By the action of CO on* OsC_4 *at* 100°C *and* 50 *atm pressure.*

$$OsO_4 + 9CO \longrightarrow Os(CO)_5 + 4CO_2$$

(b) Osmium Enneacarbonyl $Os_2(CO)_9$: It is a yellow crystaline solid and prepared by the reaction of osmium tri-iodide with carbon monoxide in the presence of copper. It is more stable towards heat than $Ru_2(CO)_9$. It melts at 224°C and sublimes without decomposition.

Carbonyls of Iridium

It forms two carbonyls.

(a) Iridium Octacarbonyl $Ir_2(CO)_8$: It is prepared by the reaction of either KIr_2Br_6 or KIr_2I_6 with carbon monoxide at 200°C and 200 atm. pressure. It is a yellow crystalline solid having m.p. 160°C. It is easily converted into the tricarbonyl by the action of heat, alkalies or acids.

(b) Iridium dodecarbonyl, $Ir_4(CO)_{12}$: It forms orange yellow rhombohedral crystals which decompose at 200°C. It is best prepared by treating iridium tri-iodide with carbon monoxide under pressure below 100°C in the presence of copper. It is chemically inert.

Carbonyls of Platinum

Preparation

(i) *When carbon monoxide is passed over platinous chloride at* 250°C, $PtCl_2.2CO$ *and* $2PtCl_2.3CO$ *are obtained. On heating these yield* $PtCl_2.CO$.

$$3PtCl_2 + 5CO \longrightarrow PtCl_2.2CO + 2PtCl_2.3CO$$

(ii) *Pure* $PtCl_2.CO$ *may be prepared by the sublimation of higher carbonyl chlorides through a tube at* 250°C,

$$PtCl_2.2CO \longrightarrow PtCl_2.CO + CO$$

$$2PtCl_2.3CO \longrightarrow 2[PtCl_2.CO] + CO$$

(iii) $PtCl_2.2CO$ *may be obtained by heating Pt foil or sponge at* 240-250°C *with carbon monoxide and chlorine.*

$$2CO + Cl_2 + Pt \longrightarrow PtCl_2.2CO$$

Properties

These carbonyls are decomposed by water and hydrochloric acid.

$$PtCl_2.CO + H_2O \longrightarrow Pt + 2HCl + CO_2$$

$$PtCl_2.2CO + H_2O \longrightarrow Pt + 2HCl + CO_2 + CO$$

$$PtCh.CO + HCl \longrightarrow H[PtCl_3.CO]$$

$$PtCh.2CO + HCl \longrightarrow H[PtCl_3.CO] + CO$$

J. Vanadium Carbonyl, $V(CO)_2$: In 1960, it was reported by *Caldrazzo* and *Alberda* that $V(CO)_6$ could be prepared by the action of VCl_3 or $VOCl_3$ on metal arnine reducing system at 80° – 150°C in an atmosphere of carbon monoxide under high pressure. They utilised magnesium and zinc as metal and pyridine as amine. When the reaction mixture was evaporated, acidified and extracted with ether, a bluish green solid separates out. Vanadium carbonyl is air sensitive and isomorphous with the carbonyls of sixth group [$Cr(CO)_6$, $Mo(CO)_6$, $W(CO)_6$].

STRUCTURES OF METAL CARBONYLS

1. Effective Atomic Number Rule

The most satisfactory formula of carbon monoxide is : C O :

From the structure of carbon monoxide, it is probable that the *lone pair of electrons on the carbon atom can be used by forming a dative bond with certain metals ($M \leftarrow C = O$). Thus, ($M \leftarrow C = O$) types of bonds were assumed to be present in metal carbonyls.*

Carbonyls	*Atomic Number of the metal*	*Number of electron contributed by CO groups*	*E.A.N.*	*Succeeding inert gas*
$Cr(CO)_6$	24	12	36	Kr(36)
$Fe(CO)_5$	26	10	36	Kr(36)
$Ni(CO)_4$	28	8	36	Kr(36)
$Mo(CO)_6$	42	12	54	Xe(54)
$Ru(CO)_5$	44	10	54	Xe(54)
$W(CO)_6$	74	12	86	Rn(86)
$Os(CO)_5$	76	10	86	Rn(86)

In the formation of M ← C = O bonds, the electrons are supplied *solely by the molecules of carbon monoxide and the metal atom is thus said to have* ***zero–valency****. The number of molecules of carbon monoxide which can unite with one atom of the metal is apparenty controlled by the tendency of the metal atom to acquire the effective atomic number of the next inert gas.* This is given for the stable monomeric carbonyl:

$$\text{E. A. N.} = m + 2y = \text{G}$$

where *m* is the atomic number of the metal *M* nd v is the number of carbon monoxide molecules in one molecule of the carbonyl $M(CO)_y$ and G is the atomic number of the next inert gas.

On the basis of EAN rule it can be explained why Ni atom fails to form a hexacarbonyl, $Ni(CO)_6$. Non–formation of $Ni(CO)_6$ is because of the fact that EAN of Ni atom in $Ni(CO)_5$ would be equal to 28 + 2 × 6 = 40 which is not the atomic number of any of the noble gases.

Mononuclear carbonyls having the metallic atom with odd atomic number: $V(CO)_6$ and hypothetical carbonyls viz $Mn(CO)_5$ and $Co(CO)_4$ are the examples of such carbonyls. These carbonyls *do not obey* EAN rule as shown below:

$V = 23e^-$	$Mn = 25e^-$	$Co = 27\ e^-$
$6CO = 12e^-$	$5CO = 10e^-$	$4CO = 8e^-$
$V\ (CO)_6 = 35e^-$	$Mn\ (CO)_5 = 35e^-$	$Co\ (CO)4 = 35e^-$

The metals with odd atomic numbers cannot form mononuclear carbonyls because they cannot acquire an inert gas electronic configuration by combination with CO molecules. Therefore, such metals form poly-nuclear carbonyls. For example, the simplest carbonyl formed by Mn(25) and Co(27) are $Mn_2(CO)_{10}$ and $Co_2\ (CO)_8$ respectively.

Evidence for the structure of polynuclear carbronyls is not complete.

Polynuclear Carbonyls

Sidgwick and Bailey have shown that the formulae of polynuclear carbonyl can be fitted to an extension of the same rule. For the carbonyl $M_x(CO)_y$ the general relation is

$$G - \frac{xm + 2y}{x} = x - 1$$

where *G* = The atomic number of the next inert gas,

m = The atomic number of metal atoms and

y = The number of CO molecules in one molecule of the carbonyl.

As regards polynuclear carbonyls such as $Mn_2(CO)_{10}$, $Co_2(CO)_9$, $Fe_2(CO)_9$, $Fe_3(CO)_{12}$ etc. obey the E.A.N. rule, their E.A.N. per atom of metal is 36. For example; E.A.N. of $Mn_2(CO)_{10}$ may be calculated as:

Electrons from 2Mn atom	$= 25 \times 2 = 50$
Electron from 10CO molecules	$= 10 \times 2 = 20$
Electrons from one Mn–Mn bond	$= 1 \times 2 = 2$
Total	= 72

E.A.N. for one Mn atom = 72/2 = 36

The formation of binuclear carbonyls having metal atoms with *odd atomic number* can also be explained on the basis of 18–electron rule as shown below for $Co_2(CO)s$ and $Mn_2(CO)_{10}$ carbonyls:

$Co_2(CO)_8$ (*Co*	$= 3d^7\ 4s^2$)		$Mn_2(CO)_{10}$ (*Mn* $= 3d_5 4s^2$)
2 *Co*	$= 2x9e^-$	$= 18e^-$	$2Mn = 2\times 7e^- = 14e^-$
8 Co	$= 2 \times 8e^-$	$= 16e^-$	$10\ CO = 10 \times 2e^- = 20e^-$
Co – Co bond $1 \times 2c^-$		$= 2e^-$	Mn – Mn bond $= 2e^-$
∴ *Electrons on one Co*		∴	*Electrons on one*
atom $= 18e^-$			*Mn atom* $= 18e^-$

Here it may be noted that although, Fe has an even atomic number (= 26), the formation of its binuclear carbonyl, $Fe_2(CO)_9$ can also be accounted for by the 18–electron rule as shown below:

$$2Fe(Fe = 3d^6\ 4s^2) = 2 \times 8e^- = 16e^-$$

$$9Co = 9 \times 2e^- = 18e^-$$

$$Fe^-\ Fe\ bond = 1 \times 2e^- = 2e^-$$

∴ *Electrons on one Fe atom* $= 36e^-/2 = 18e^-$

Objection

From the X–ray or electron diffraction method, it is well established that the bonds are intermediate between the M – C ≡ O and M = C = O states, *i.e.,* there is some double bond character in M – CO.

Thus, the above older views cannot explain the double bond character in M–CO. This has been explained by both molecular orbital as well as by valence bond method.

2. Molecular Orbital Approach

This can be understood by considering the following facts :

(a) Carbon undergoes *sp* hybridisation involving 2*s* and 2*px* orbitals, forming two hybrid orbitals, which are designated as SP_a and SP_b. These orbitals are oriented through 180°. Now we can write the configurations of C and O atoms as follows:

$$C[(1s)^2\ (2SP_a)^2\ (2SP_b)\ (2Py)^1\ (2P_z)^0]$$

$$O[(1s)^2\ (2s)^2\ (2P_x)^1\ (2p_y)^1\ (2p_z)^2]$$

According to the molecular orbital theory, carbon and oxygen atoms undergo overlapping to form bonds in CO as follows:

(i) $2SP_b$ *hybrid orbital of carbon and* $2p_x$ *of oxygen overlap to form a localised σ bond.*

(ii) $2p_y$ *of carbon and* $2p_y$ *of oxygen overlap to form α π-bond.*

(iii) $2p_z$ *of carbon and* $2p_z$ *of oxygen overlap to form another π-bond.*

The distribution of electrons of carbon and oxygen (of the second quantum shell) in the various molecular orbitals formed is as follows:

(i) *There, will be 2 electrons in σ bonding orbital.*

(ii) *There will be 2 electrons in the π bonding orbital which is formed by the combination of* $2p_y$ *atomic orbitals of carbon and oxygen.*

(iii) *There will be 2 electrons in the π–bonding orbital, formed by the combination of* $2p_z$ *orbitals of oxygen and carbon.*

(iv) *There will be 2 non–bonding electrons in the* $2SP_a$ *hybrid orbital of carbon.*

(v) *There will be 2 non–bonding electrons in the 2s atomic orbital of oxygen.*

(vi) *There will be no electron in the antibonding molecular orbitals, formed as result of σ and π overlappings.*

As the total number of bonding electrons is six and that of antibonding electrons nil, bond order of the molecule is three. Hence, the number of bonds between carbon and oxygen atoms in CO molecule is 3, one σ and two π.

The lone pair of electrons on carbon could be expected to form a strong dative bond (a) due to the electron density remaining close to the nucleus of the carbon atom. As metal atom-carbon monoxide bonds are readily formed in metal carbonyls, it is expected that there is some additional bonding mechanism in the formation of metal–carbon monoxide bonds in the metal carbonyls. The accepted view is that the weak M–C a is formed along with M–C π bond.

The above situation can be understood in the following way :

(i) Firstly, there is a dative overlapping of filled carbon o orbital, *i.e.*, $2SP_a$ hybrid orbital with an empty metal σ orbital $(M \overset{\sigma}{\leftarrow} CO)$.

(ii) Secondly, there is a dative overlapping of a filled d-orbitals of metal with empty antibonding $-p_\pi$ orbital of the carbon atom $(M \rightarrow CO)$ resulting in the formation of a dative π bond.

$$\text{metal} \underset{\text{Dative } \pi \text{ bond}}{\overset{\text{Dative } \sigma \text{ bond}}{\rightleftarrows}} \text{Carbon monoxide}$$

Shaded portions in Figs. (2.8 and 2.9) indicate the filled orbitals whereas empty portions indicate vacant orbitals, *i.e.*, having no electrons.

As there is a drift of metal electrons into carbon monoxide $(M \overset{\pi}{\rightarrow} CO)$, orbitals will tend to make the CO as a whole negative and at the same time there is the drift of electrons from CO to the metal $(M \overset{\sigma}{\leftarrow} CO)$ to make the CO positive, thus enhancing the acceptor strength of the π orbitals. Thus, in a way the formation of a bond strengthens the π bond formation and *vice versa.*

The strongest support for this type of bonding mechanism comes from IR spectra. A major consequence of back donation of electrons into π* orbitals of CO molecule is that the carbon-oxygen bond order in them, which is three in the free uncomplexed CO molecules, is reduced. This reduction in bond order on complexation reduces the force constant enabling the bond to be stretched with less energy.

A decrease in the C–O stretch frequency value is thus predicted. It is observed that C–O stretch frequency in the uncomplexed CO molecules is 2155 cm^{-1} while in a complexed molecule, it has been observed to be in the neighbourhood of 2000 cm^{-1}. This observation, hence, supports the mechanism proposed.

Dinuclear carbonyls: (a) *$Mn_2(CO)_{10}$ $Tc_2(CO)_{10}$ and* $Re_2(CO)_{10}$ *carbonyls.* In all these carbonyls each M atom gets directly linked with the other M atom by a σ–bond (M–M σ–bond) and five terminal carbonyl groups ($-C \equiv O$) by a co-ordinate bond ($M \leftarrow C \equiv O$). The presence of M–M bond is also supported by the diamagnetic nature of these carbonyls. These carbonyls are not having bridging carbonyl groups ($>C = O$) in between M atoms.

As the coordination number of each M atom in these carbonyls is six, each M atom is d^2sp^3 hybridised. All the electrons in these carbonyls are paired and hence these carbonyls are diamagnetic. The structure of $Mn_2(CO)_{10}$ has been explained in detail elsewhere.

(*b*) $Co_2(CO)_9$, *$Fe_2(CO)_9$ and $Os_2(CO)_9$ carbonyls*: These carbonyls have both types of carbonyl groups which are bridging carbonyl groups ($>C = O$) and terminal carbonyl groups ($-C \equiv O$). Bridging carbonyl groups exist between two metal atoms. The number of bridging carbonyl groups present in $Co_2(CO)_9$ (only in one isomeric form), $Fe_2(CO)_9$ and $Os_2(CO)_9$ has been 2, 3 and 1 respectively while the number of terminal carbonyl groups is 6, 6 and 8 respectively.

Each of the bridging carbonyl group gets attached with the metal (M) by bond while each of the terminal carbonyl groups is attached with the metal atom by $O \equiv C \rightarrow M$ coordinate bond. All these carbonyls have M–M bond. The presence of M–M bond confirms the diamagnetic nature of these carbonyls.

Trinuclear carbonyls: Here an attempt is made to consider the structures of $Fe_3(CO)_{12}$, $Ru_3(CO)_{12}$ and $Os_3(CO)_{12}$ carbonyls. The trinuclear corbonyls viz $Os_3(CO)_{12}$ and $Ru_3(CO)_{12}$ possess similar structures whereas $Fe_3(CO)_{12}$ has a different structure. $Os_3(CO)_{12}$ and $Ru_3(CO)_{12}$ contain three metal (M) atoms. Each of these gets linked with the remaining two M atoms. Thus, there are three M–M bonds in each of these carbonyls. All the three M–M bonds are situated in an equilateral triangle. Each M atom gets attached with four terminal CO groups which make an approximately octahedral configuration around each M atom. These molecules do not have any bridging CO groups.

Evidence for the structures of the polynuclear carbonyls is still incomplete. In most of the poly nuclear complexes the coordination centres are joined through ligand (CO) bridges but such crystal structure determinations have been carried out that have shown that direct metal metal bonds are an important features of the polynuclear carbonyls.

3. Valence Bond Method: In this method, the molecule may be represented by resonance structures.

$$\bar{M} - \bar{C} - O \longleftrightarrow M = C = O$$

With a large amount of the double bond character, it is this structure that accounts for their stability. From either the molecular orbital or the valency bond view point, *back donation is seen in both.*

We shall now discuss the structures of some carbonyls taking the older view as well as the modern view into consideration. First of all, We shall discuss the monocarbonyls and later the polynuclear carbonyls.

Structure of Ni(CO)$_4$

(i) *The vapour density of nickel carbonyl and the freezing points of its solution in benzene indicate the molecular formula to be Ni (CO)$_4$.*

(ii) *Electron diffraction studies made on this compound in the vapour state and X–ray diffraction studies made on this compound in the solid state have shown that* $Ni(CO)_4$ *molecule has tetrahedral shape with Ni–C–O linear units. The Ni – C bond length in this molecule is* 1.50 Å *which is shorter by* 0.32 Å *in comparison to Ni–C single bond length* (= 1.82 Å) *found in carbonyls. The C–O bond length in this carbonyl has been found to be equal to* 1.15 Å *which is larger than the C–O bond length in CO molecule* (= 1.128 Å).

(iii) *Raman spectral studies have revealed that the oxygen atom is triply bonded to the carbon atom in the CO group Thus the nickel atom in the nickel carbonyl must be tetrahedrally hybridised.*

Old Orbital Diagram: The electronic configuration of nickel atom is given as:

$$Ni_{28} = 1s^2 2s^2 2p^6\ 3s^2 3p^6 3d^8 4s^2$$

Previously, it was thought that the above configuration of nickel reverted to the energetically similar $3d^{10}$ outer arrangement permitting the hybridisation $4s^1 4p^3$. Thus, four unoccupied orbitals in the vlalency shell were available which could accept four carbon monoxide groups. Hence, it gives rise to the tetrahedral configuration. This orbital diagram could not be supported by experimental facts.

Tetrahedral shape of $Ni(CO)_4$ arises due to sp^3 hybridisation of Ni–atom. As $Ni(CO)_4$ molecule is diamagnetic, all the ten electrons present in the valence shell of Ni–atom (Ni = $3d^8$ $4s^2$) are paired in 3d orbitals. Thus the valence–shell configuration of Ni–atom in $Ni(CO)_4$ molecule becomes $3d^{10}$. $4s^{\circ}$. CO → Ni–bond is caused by the overlap between the empty sp^3 hybrid orbital on No–atom and doubly–filled sp hybrid orbital On C atom in CO molecule.

Because of the formation of four OC → M bonds, a large negative charge gets accumulated on the central Ni atom. This is most unilikely. In such a situation, Pauling suggested that the double bonding occurs with the back donation of d-electron from Ni atom to CO ligands to such an extent that electroneutrality principle is obeyed. According to this principle the electron pair is not shared equally between Ni and C–atoms of CO ligands but gets attracted more strongly by C–atom which prevents the accumulation of negative charge on Ni–atom and is in keeping with the greater electronegativity of C – atom compared to Ni atom (C = 2.5, Ni = 1.8).

Evidences: The bond structure shown in Fig.2.13B is supported by the following reactions:

(i) *When an alcoholic solution of the carbonyl is treated with ortho-phenanthroline to yield a stable ruby–red compound* $Ni(CO)_2$ *phen, it confirms that the two C = O groups of Ni(CO)4 are replaced by one molecule of phenanthroline.*

(ii) *Similarly, the reaction of* $Ni(CO)_4$ *with diarsine indicates the two C ≡ O groups*

2. Structure of $Fe(CO)_5$: The various evidences are:

(i) *The vapour density and the freezing points of benzene that its molecular formula is* $Fe(CO)_5$.

(ii) *Electron diffraction, Roman and infra-red studies of this compound have shown that* $Fe(CO)_5$ *molecule has trigonal bipyrmidal shape. Electron diffraction study of* $Fe(CO)_5$ *in the vapour state has shown that Fe–C axial bond and Fe–C basal bond lengths are equal to* 1.797 Å *and* 1.842 Å *respectively while X–ray study has shown that these bond lengths are almost the same. Trigonal bipyramidal shape of* Fe(CC)$_5$ *results from* dsp^3 *hybridisation of Fe–atom.*

Thus, five empty orbitals are available which can accept five carbon monoxide groups, giving rise to a trigonal bipyramid structure. However, this can not be supported by the experimental studies.

In this molecule Fe is in zero oxidation state and since $Fe(CO)_5$ is diamagnetic, all the eight electrons in the valence–shell of Fe atom (Fe = $3d^6$, $4s^2$) get paired in $3d$ orbitals. Thus the valence–shell configuration of Fe–atom in $Fe(CO)_5$ becomes $3d^8$. $4s^o$. CO $\rightarrow$ Fe σ–bond results by the overlap between empty dsp^3 hybrid orbital on Fe–atom and doubly–filled sp hybrid orbital on C–atom in CO molecule.

NOBLE GASES, THEIR COMPOUNDS AND CLATHRATES

The noble gases *helium, neon, argon, krypton, xenon* and *radon* are the elements present in group zero. All these gases, except radon, are present in the atmosphere. Radon results from the disintegration of radium and is radioactive itself.

The atomic numbers of the elements He, Ne, Ar, Kr, Xe and Rn are 2, 10, 18, 36, 54 and 86. respectively. Their electronic configurations. All these elements except helium, arc having the *highly stable* s^2p^6 configuration in the outer (valency) shell. Helium is having the stable s^2 configuration. Because of the highly stable s^2p^6 configuration in the, outermost shell, these gases are having very little tendency to form chemical compounds with other elements. Actually, till middle of this century no chemical reaction of these elements was known and they were re-known as *inert gases*.

However, now researches have shown that they do enter into chemical reactions with some specific elements under certain specific conditions to form chemical compounds. The word, 'Inert', therefore, has been no longer considered appropriate.

It is more usual now to call them as noble gases, the word *'noble'* signifying that they enter into very few chemical reactions in the same way that the noble metals are often reluctant to react and are the least reactive metals.

COMPOUNDS FORMATION BY INERT GASES

Earlier attempts to prepare any compound of rare gases were met with failure. The main reasons for this are:

(i) *The complete pairing of all electrons present and the absence of any bonding orbitals.*

(ii) *The existence of these atoms in stable energy states as shown in the Pauling type energy level diagram of these elements.*

Compound Formation Under Excited Conditions

If we unpair the $1s^2$ electron in the helium atom and promote it to 2s state, then a condition of chemical activity is expected. But this process requires some 1932 kJ of energy per gm atom and can only be realised spectroscopically or under conditions of electric discharge of electron bombardment.

$$He = 1s^2 \text{ (normal, inactive)}$$

$$He = 1s^1 2s^1 \text{ (excited, active).}$$

In actual practice, some compound formation under excited conditions' takes places. Some well established examples are:

(i) *In a discharge tube,* He_2^+, HeH^+ *and* HeH^{2+} *have been reported but these survive only momentarily.*

(ii) *In a glow discharge, a comparative stable mercury halide* ($HgHe_{10}$) *has been reported to be formed.*

(iii) *When tungsten is bombarded with electrons in an atmosphere of helium, tungsten halide* WHe_2 *is obtained.*

(iv) *Unstable halides of elements such as bismuth, thallium, indium, zinc, sodium, potassium, etc. have been reported to be formed.*

(v) *Metal electrodes in discharge tubes absorb inert gases to form* Pt_3He, *FeHe and FeAr which are probably the interstitial compounds.*

The compound Pt_3He has been thoroughly studied and is found to be different from platinum metal 1 in the following properties:

(i) Pt_3 He *has a ixed decomposition temperature.*

(ii) Pt_3 He gives *an X–ray diffraction pattern which is characteristic of an amorphous substance.*

(iii) Pt_3He *possesses a different density, solubility and electrical resistance from platinum.*

(iv) Pt_3He *is not decomposed by aqua-regia whereas Pt is attacked by it.*

Compound Formation by Physical Trapping

Until recently, the only well established compounds of the noble gases were those in which gases were trapped in the crystal lattices of other compounds. *These compounds are termed by Powell as* cage compounds or clathrate compounds. *These are so called because one component of the compounds is trapped within the cavities of the crystal lattices of the other component and there is no chemical bond formation between the two components.* These compounds are decomposed on melting or on dissolution of the compounds in any suitable solvent. Examples are:

(i) If quinol (1 : 4 dihydroxy benzene) is crystallised in the presence of heavier noble gases under a pressure of 10 – 40 atmospheres, the gas is trapped in cavities of about 4 Å diameter in the structure. The gas escapes only when the clathrate is dissolved, only Ar, Kr and Xe form clathrate compounds.

Helium and neon fail to form quinol clathrates presumably due to the small sizes of their atoms which enable them to escape from the quinol lattice. Powell confirmed these clathrate compounds of argon, krypton and xenon with quinol of the type:

1 molecule of noble gas: 3 molecules of quinol.

(ii) Similarly, the gases Ar, Kr, and Xe may be trapped in cavities when water solidifies. The so called inert gas hydrates have formulae approximating to $5H_2O$: 1. gas atom.

It is highly probable that clathrates are not chemical compounds.

Compound Formation by Dipole-induced Dipole Attraction

The noble gas atoms may be polarised by a strong dipole on other molecules and possess a weak induced dipole of their own. The compounds are expected to be unstable due to weak forces. The solubility of the noble gases in water is high and increases with increased size. It is probably due to the reason that larger atoms are more easily polarised by the surrounding water molecules.

The existence of compounds of argon and krypton with water was confirmed by *Tammann* and *Krige* who formulated them as hexahydrates. They also reported $Kr.6D_2O$ and

Tammann and *Krig* prepared the compounds by compressing the inert gases (Ar, Kr and He) with ordinary water or heavy water at low temperature, and pressure is released. These compounds are called hydrates. The compounds such as $Ar.6H_2O$, $Kr.6H_2O$ and $Xe.6H_2O$ have been obtained and studied.

Hydrates of inert gases are crystalline compounds and their stabilities decrease rapidly from the radon compound down through to argon compound. Similar types of compounds are formed with phenols having the formulae:

$$Kr.2C_6H_5OH$$

$$Xe.2C_5H_5OH.$$

The compounds of inert gases with phenol are isomorphous with the hydrogen sulphide derivative such as $H_2S.\ 2C_6H_5OH$. It is important to remark that lithium ions possess tendency to form clusters with inert gases of the type:

$$Li^+nG(n = 1 \text{or } 2) \text{ where } G = \text{Inert gas}$$

The stabilities of such aggregates and the quantities of gas held increase with the increasing size of the inert gas atoms.

Some Recently Discovered Inert Gas Compounds

Only krypton and xenon have been shown to form true chemical compounds. Neither of them were discovered before 1962. Their chemistry still remains limited because the noble gas atoms are only bonded to either fluorine or oxygen in all their stable compounds.

Bartlett, in 1962, found that oxygen gas reacts with PtFe to give a solid compound of the formula O_2PtF_6. The X–ray examination revealed that the compound actually is made up of O_2^+ and PtF_6^- ions and could be represented as $O_2^+PtF_6^-$.

As the ionisation energy of xenon (1176 *kJ*/mol) *is very close to that of oxygen* (1320 *kJ/mol*) *it occurred to Bartlett that the reaction between xenon and* PtFe *should be tried.* He actually found that the reaction took place readily even to room temperature to yield a red solid compound of the composition $XePtF_6$. *This was the first real compound of any of the noble gases.*

The reaction has since been shown to be more complicated, and the product is really $[XeF]^+\ [Pt_2F_{11}]^-$.

$$Xe[PtF_6] + PtF_6 \xrightarrow{25°C} XeF]^+[PtF_6]^- + PtF_5$$

$$\xrightarrow{\text{heat } 60°C} [XeF]^+[Pt_2F_{11}]^-.$$

Soon after this it was found that Xe and F_2 reacted at 400°C to give a colourless volatile solid XeF_4. This has the same number of valency electrons as, and is isostructural with, the polyhalide ion $[ICl4]^-$. Following these discoveries there was rapid extension of the chemistry of the noble gases, and in particular of xenon.

The reaction of xenon with PtF_6 demonstrated for the first time that xenon is not a totally inert gas. This led to other investigations in the same direction.

From the ionisation energy values of these elements , it is seen that except radon, which is unstable, only xenon and, to some extent krypton, has ionisation energy which is comparable with common elements such as carbon, nitrogen and oxygen. Xenon thus forms the largest number of compounds. Krypton comes next.

Neon, argon and helium, which have much higher ionisation energies form hardly any compound. Moreover, since reactivity of xenon and krypton is because of their tendency to part with their electrons, it is quite clear that the combining atom should have *strong affinity for electrons, i.e., should be strongly electronegative.* It is due to this that xenon and krypton form compounds mostly with strong electron acceptors, viz., fluorine and oxygen. The most stable of these compounds are *fluorides, oxyfluorides* and *oxides.*

Xenon reacts directly with fluorine. Its compounds with oxygen are obtained.

Why noble gas compounds were discovered very late?

The main reason for this is their stable electronic configuration (ns^2 *or* ns^2p^6) which was regarded as the criterion of chemical inertness. Although existence of compounds having more than eight valency electrons were soon confirmed *e.g.,* $PCl_5(10)$, $SF_6(12)$, IF_7 (14), etc. no serious attempt was made to over–rule the stable octet as the criterion of inertness in noble gas. Some other factors responsible for the late discovery of noble gas compounds are as follows:

(i) Good quality fluorine was not available.

(ii) Fluorine–resistant greases were not available.

(iii) There arises difficulty in handling fluorine gas (non– availability of inert materials to handle fluorine gas).

(iv) Noble gases are not available in plentiful quantities.

Why do noble gases form compounds only with fluorine and oxygen?

As the chemical reactivity of noble gases involves the loss of electrons, the highly electronegative element (F and O) and group (*e.g.*, $OSeF_5$ and $OTeF_5$ which contain both oxygen and fluorine can) combine with gases very readily.

Xe reacts directly only with F_2. However, oxygen compounds can be obtained from the fluorides. There is some evidence for the existence of $XeCl_2$ and $XeCl_4$, and one compound is known with a Xe–N bond. Hence the important noble gas compounds include fluorides, oxyfluoride and oxides of xenon and krypton.

Compounds of Xenon

Xenon and fluorine react under a variety of conditions like temperature, pressure mixing ratio of the constituents, ultraviolet irradiation, electric discharge, etc., to produce xenon fluorides.

For examples:

$$\underset{(2:1 \text{ ratio})}{Xe + F_2} \xrightarrow{400°C, 2Hr.} XeFe_2$$

$$\underset{(1:5 \text{ ratio})}{Xe + 2Fe_2} \xrightarrow{450°C, 475°C} XeFe_4$$

$$\underset{(1:20 \text{ ratio})}{Xe + 3F_2} \xrightarrow[50 \text{ atm.}]{300°C, 10-20 \text{ hr.}} XeF_6.$$

It can be seen from the pressure versus temperature diagrams of xenon fluorides that XeF_4 will always be contaminated with some amounts of lower (*i.e.*, XeF_2) and higher (*i.e.*, XeF_6) fluorides. It can be purified by making use of its inability to donate F^- ion to AsF_5 acceptor.

Impure Xep_4 is dissolved in BrF_5 and reacted with an excess of AsF_5. Xenon difluoride and xenon hexafluoride form involatile salts while unreacted BrF_5 is distilled off at 273°C. The unchanged pure XeF_4 is then obtained by vacuum distillation at room temperature.

Why xenon and bromine, argon and fluorine do not form compounds like xenon fluorides?

Electronegativity of bromine is much less than that of fluorine, so bromine is unable to bring about promotion of 5p electrons of xenon to 5d–orbitals. As the electrons remain paired, no compounds are formed. Ionization energy of argon is quite high, *i.e.,* higher than that of xenon. Therefore, the energy released by Ar–F bonds is insufficient to promote electron from 4p to 4d electrons to give rise to unpaired electrons. As the electrons remain paired, no compounds are formed.

Preparation

(i) It has been prepared by irradiating a mixture of xenon and fluorine with light from a high pressure mercury arc.

$$Xe + F_2 \xrightarrow[\text{Mercury lamp}]{\text{Light}} XeF_2$$

(ii) It may be conveniently prepared by circulating a mixture of Xe and' fluorine through a nickel tube at 400°C for about 8 hours.

$$Xe + F_2 \xrightarrow{400°C} XeF_2$$

When the mixture of XeF_2 unreacted Xe and fluorine is fed into a U–tube maintained at – 50°C. XeF_2 is condensed into it.

XeF_2 has to be removed as soon as it is formed otherwise it reacts with more fluorine to form XeF_4. The XeF_2 can be condensed and thus removed from the unreacted xenon and fluorine by passing the reaction product through a U tube maintained at – 50°C.

$$XeF_2 + F_2 \longrightarrow XeF_4$$

(iii) It can be prepared by the reaction of xenon with O_2F_2 at –118°C.

$$Xe + O_2F_2 \xrightarrow{-118°C} XeF_2 + O_2$$

Properties

(i) It forms transparent crystalline solid melting at –127°C.

(ii) It is less volatile than hexafluoride and can be stored in nickel containers for a long time when it is free from impurities.

(iii) It dissolves in anhydrous liquid HF and the solution does not conduct electricity,

(iv) XeF_2 is a mild fluorinating agent. At room temperature, ethylene is fluorinated to form 1,1 and 1,2 – difluoroethanes.

$$XeF_2 + 2C_2H_4 \longrightarrow \begin{matrix} CH_2F \\ | \\ CH_2F \end{matrix} + \begin{matrix} CH_2F \\ | \\ CH_3 \end{matrix} + 2He$$

With propylene, the main product is, 1,1–difluoropropane

$$XeF_2 + C_3H_6 \longrightarrow CF_2HCH_2CH_3 + Xe$$

(v) XeF_2 reacts with oxides of nitrogen. With NO_2 the reaction is slow but with nitric oxide it is rapid and is represented as:

$$XeF_2 + 2NO \longrightarrow 2NOF + Xe$$

(vi) XeF_2 reacts quantitatively with hydrogen as follows:

$$XeF_2 + H_2 \longrightarrow 2HF + Xe$$

(vii) Its hydrolysis proceeds very rapidly in an alkaline medium but slowly in neutral or acid medium. When it is boiled with 1 molar solution of sodium hydroxide, it undergoes hydrolysis liberating oxygen quantitatively according to the following equation:

$$XeF_2 + H_2O \longrightarrow Xe + \frac{1}{2}O_2 + 2HF$$

At first the hydrolysis produces a bright yellow colour which rapidly disappears and the solution becomes colourless at room temperature.

(viii) XeF_2 forms addition compounds with SbF_5, TaF_5, NbF_5) RuF_5 etc.

$$XeF_2 + 2SbF_5 \longrightarrow XeF_2.2SbF_5$$

(ix) On heating it undergoes disproportionation.

$$2XeF_2 \longrightarrow Xe + XeF_4$$

XeF_2 oxidises Cl^- to Cl_2, I^- to I_2 and cerium (III) to cerium (IV).

$$XeF_2 + 2HCl \longrightarrow 2HF + Xe + Cl_2$$

$$SO_4^{2-} + XeF_2 + Ce_2^{III}(SO)_3 \longrightarrow 2Ce^{IV}(SO4)_3 + Xe + F_2$$

Xenon Tetrafluoride

Preparation

(i) It is prepared by heating a mixture of fluorine and xenon in the ratio of 5 : 1 in a nickel vessel at 450 – 475°C and 6 atmospheric pressure for a few hours. After heating, the reaction is quenched by cooling the nickel vessel rapidly in cold water. The excess of fluorine is removed by pumping at – 78°C.

$$Xe + 2F_2 \xrightarrow{450-475°C} XeF_4$$

The tetrafluoride prepared under the above mentioned conditions contains less than 8% XeF_2 and XeF_6 as impurities. These impurities can be removed from XeF_4 by sublimation.

(ii) It is also prepared by passing an electric discharge through a mixture of xenon and fluorine.

Properties

(i) At room temperature, it forms transparent crystals having m.p.– 114°C.

(ii) It is quite unstable and can be stored in a nickel vessel.

(iii) It is less volatile than hexafluoride.

(iv) It dissolves in trifluoroactic acid.

(v) It is hydrolysed by H_2O fumming XeO_3

$$3XeF_4 + 6H_2O \longrightarrow XeO_3 + 2Xe + \frac{3}{2}O_2 + 12HF$$

This reaction is indeed a major hazard in Xe/F chemistry, since XeO_3 is highly explosive, and the complete exclusion of moisture is therefore essential. Interestingly the maximum yield of XeO_3 is 33% rather than the 50% that would be expected from a simple disproportionation of and the following reaction sequence has been suggested to explain this:

$$3Xe^{IV}F_4 + 6H_2O \longrightarrow \{Xe^{II}O\} + \{Xe^{VIII}O_4\} + 12HF$$

$$\{Xe^{II}O\} \xrightarrow{\text{Decomposition}} 2Xe^0 + O_2$$

$$\{Xe^{VIII}O_4\} \xrightarrow{\text{Decomposition}} X^{VI}O_3 + \frac{1}{2}O_2$$

The stoichiometry of the reaction also depends sensitively on the precise conditions of hydrolysis,

(vi) It reacts with NO, NO_2, HCl and NH_3.

$$XeF_4 + 4NO \longrightarrow 4ONF + Xe$$

$$XeF_4 + 4NO_2 \longrightarrow 4O_2NF + Xe$$

$$XeF_4 + 4HCl(g) \longrightarrow 2HF + 2Cl_2 + Xe$$

$$XeF_4 + 6NH_3(g) \longrightarrow 4NH4F + N_2 + Xe + H_2$$

(vii) XeF_4 is a strong fluorinating agent. It fluorinates ethylene and propylene.

$$2C_2H_4 + XeF_4 \longrightarrow \begin{array}{c} CH_2F \\ | \\ CH_2F \end{array} + \begin{array}{c} CH_2F \\ | \\ CH_3F \end{array} + Xe$$

It fluorinates compounds:

$$XeF_4 + 2SF_4 \longrightarrow Xe + 2SF_6$$

$$XeF_4 + Pt \longrightarrow Xe + PtF_4$$

$$XeF_4 + C_6H_6 \longrightarrow Xe + C_6H_5F + HF$$

(viii) When XeF_4 is warmed with excess of xenon, XeF_2 is formed

$$XeF_4 + Xe \longrightarrow 2XeF_2$$

(ix) With platinum, XeF_4 in anhydrous HF gives PtF_4

$$2Pt + XeF_4 \longrightarrow PtF_4 + 2Xe$$

(x) There is some evidence that XeF_4 forms an addition compound of composition $XeF_4.2SbF_5$.

Xenon Hexafluoride (XeF_6)

Preparation

(i) It is very difficult to prepare. However, it is prepared by heating a mixture of xenon and fluorine in the ratio of 1: 20 in a nickel vessel and at 300°C and 50 atmospheric pressure.

$$Xe + 3Fe_2 \xrightarrow[50\ atm.]{300°C} XeF_6$$

The yield of this method is 95%. It becomes 100% when the pressure is. 2 atmospheres and the temperature 700°C.

(ii) It is prepared by the electric discharge in a mixture of fluorine and xenon in 3 : 1 ratio at low temperatures.

(iii) XeF_6 is also formed by the reaction of XeF_4 with O_2F_2 at temperatures between –133 and – 78°C

$$XeF_4 + O_2F_2 \longrightarrow O_2 + XeF_6$$

XeF_6 cannot be handled in glass or quartz apparatus because of stepwise reaction which finally produces the dangerous XeO_3:

$$2XeF_6 + SiO_2 \longrightarrow 2XeOF_4 + SiF_4$$

$$2XeOF_4 + SiO_2 \longrightarrow 2XeO_2F_2 + SiF_4$$

$$2XeO_2F_2 + SiO_2 \longrightarrow 2XeO_3 + SiF_4$$

Properties

It is a colourless crystalline solid at room temperature and melts at 47°C to give a yellow liquid but does not possess definite melting point.

(ii) It is best stored in nickel containers because it rapidly attacks glass and silica,

(iii) It is soluble in anhydrous hydrogen fluoride and is appreciably ionized as

$$HF + XeF_6 \longrightarrow XeF_5^+ + HF_2^-$$

It also reacts with alkali metal fiuorides except LiF to form heptafluorides.

$$XeF_6 + RbF \longrightarrow RbXeF_7$$

The rubidium and caesium salts are unstable and undergo decomposition.

$$2RbXeF_7 \longrightarrow XeF_6 + Rb_2XeF_8$$

(iv) Like other fiuorides, it also gets reduced by hydrogen to xenon and hydrogen fluoride.

$$XeF_6 + 3H_2 \longrightarrow Xe + 6HF$$

(v) It rapidly reacts with HCl and ammonia in accordance with the following equations:

$$XeF_6 + 6HCl \longrightarrow Xe + 3Cl_2 + 6HF$$

$$XeF_6 + 8NH_3 \longrightarrow Xe + 6NH_4F + N_2$$

(vi) When it is treated with water, it undergoes hydrolysis to form XeO_3 or $XeOF_4$.

$$XeF_6 + 3H_2O \longrightarrow XeO_3 + 6HF$$

$$XeF_6 + H_2O \longrightarrow XeOF_4 + 2HF$$

Probably this reaction proceed *via* the intermediate product $XeOF_4$. But this is still not certain.

(vii) On strong heating, it decomposes to give XeF_2 and XeF_4.

(viii) Out of three fluorides of xenon, XeFe is the best fluoronating agent.

(ix) Many addition compounds with fluorides of antimony, tin, arsenic, germanium, etc., are known.

$$AsF_5 + XeF_6 \longrightarrow XeF_6 \cdot AsF_5 \quad \text{or } [XeF_5]+ [AsF_6]^-$$

$$SnF_4 + 4XeF_6 \longrightarrow 4XeF_6 . SnF_4$$

$$XeF_6 + BF_3 \longrightarrow XeF_6 . BF_3 \text{ or } [XeF_5]^+ [BF_4]^-$$

(x) It reacts with fluoride ion donors to from $[XeFs]^{2-}$ species. For instance.

$$CsF + XeF_6 \longrightarrow CsF.XeF_6 \xrightarrow{CsF} Cs_2XeF_8 \text{ or } 2Cs^+[XeF_8]^{2-}$$

(xi) It reacts with sodium hydroxide to form white crystalline sodium perxenate.

$$XeF_6 + 1\ ONaOH \longrightarrow Na_4XeO_6 + 6NaF + 4H_2O + H_2$$

Other Fluorides

1. XeF : Xenon monofluoride was detected by electron-spin resonance spectrum in crystals of XeF_4 which is irradiated with X–rays.

2. Xenon octafluoride XeF_8 **:** Workers at the Nuclear Institute of **Jozef Stefan** in Yugoslavia studied the reaction of xenon with fluorine at 620° and pressures approaching 200 atm. They isolated yellow product at – 195°C which was volatile at – 8°C. On hydrolysis, with dilute alkali a yellow solution was obtained. Volume of fluorine indicated a Xe:F ratio of about 1.8. The above work of Yugoslavian scientists could not be confirmed in other laboratories.

Xenon Oxy Fluorides

Xenon forms two oxytetra–fluorides: (i) XeC^-_4 and (ii) XeO_2F_2. These oxy fluorides have been well characterised.

(a) Xenon Oxytetrafluoride ($XeOF_4$)

Preparation

(i) It is prepared by the controlled hydrolysis of XeFe with equimolar quantity of water.

$$XeF_6 + H_2O \longrightarrow XeOF_4 + 2HF$$

(ii) $XeOF_4$ may also be prepared by heating XeFe in a silica tube at 50°C. When the yellow colour of XeFe disappears the bulb is immediately cooled with solid CO_2. The vapours of SiF_4 are pumped out and $XeOF_4$ is collected.

$$XeF_4 + SiO_2 \longrightarrow XeOF_4 + SiF_4$$

When the yellow colour of XeFe disappears, the reaction should be stopped by cooling the tube with solid carbon dioxide otherwise $XeOF_4$ will further react with silica to form trioxide.

$$XeOF_4 + SiO_2 \longrightarrow XeO_3 + SiF_4$$

Properties

(i) It is a clear, colourless mobile liquid which can be stored unchanged in nickel containers for a long period.

(ii) It reacts with hydrogen to form xenon gas at 200°C.

$$XeOF_4 + 2H_2 \xrightarrow{200°C} Xe + H_2O + 2HF$$

(iii) With silica, it results in the formation of explosive trioxide.

$$XeOF_4 + SiO_2 \longrightarrow SiF_4 + XeO_3$$

(iv) With water it undergoes hydrolysis forming first the intermediate XeO_2F_2 and finally the explosive oxide XeO_3.

$$XeOF_4 + H_2O \longrightarrow 2HF + XeO_2F_2$$

$$XeO_2F_2 + H_2O \longrightarrow 2HF + 2XeO_3$$

(v) It forms a number of addition compounds by combination with appropriate reactants.

$$CsF.XeOF_4; \; 5RbF.2XeOF_4; \; 6KF.XeOF_4$$

(b) Xenon Dioxydifluoride ($XeChF_2$)

Preparation

Xenon trioxide reacts with xenon oxytetrafluoride to form xenon dioxide difluoride.

$$XeO_3 + XeOF_4 \longrightarrow 2XeO_2F_2$$

Properties

(i) XeO_2F_2 forms colourless crystals which melt at 30.8°C giving a colourless liquid.

(ii) With moisture it undergoes hydrolysis to form XeO_3.

$$XeO_2F_2 + H_2O \longrightarrow 2HF + XeO_3.$$

Xenon dioxide difluoride should be handled with some precautions due to the danger of explosion.

A large number of species $XeOF_4^+$, $XeOF_2^+$ $XeOF_3^+$ $XeOF^+$ and $XeO_2F_2^+$ have been identified in the course of mass spectrometric studies of noble gas compounds.

(c) Oxides of Xenon

(a) Xenon Trioxide : The direct combination of xenon and oxygen does not form any compound. However, it can be prepared by evaporation of solution produced by hydrolysis of XeF_4.

$$6XeF_4 + 12H_2O \longrightarrow 2XeO_3 + 4Xe + 3O_2 + 24HF$$

Xenon trioxide may be prepared by hydrolysis of XeF_6. The overall stoichiometry of the reaction is given by the equation:

$$XeF_6 + 3H_2O \longrightarrow XeO_3 + 6HF.$$

This reaction apparently proceeds in a stepwise fashion:

$$XeF_6 + H_2O \longrightarrow XeOF_4 + 2HF$$

$$XeOF_4 + H_2O \longrightarrow XeO_2F_2 + 2HF$$

$$XeO_2F_2 + H_2O \longrightarrow XeO_3 + 2HF$$

Properties

(i) XeO_3 is a transparent crystalline compound which is explosive in nature.

(ii) The crystals of XeO_3 are stable for days in dry air, but in humid air they deliquesce.

(iii) Its explosive nature can be realised that 200 mg of it destroys nickel container in which it was kept.

(iv) In aqueous solution, it behaves as a weak acid In basic solution, it gets hydrolysed to perxenate.

$$XeO_3 + 3H_2O \longrightarrow H^+(aq.) + H_5XeO_6^-$$

$$2XeO_3 + 4NaOH + 6H_2O \longrightarrow Xe + O_2 + Na_4XeO_6.8H_2O$$

(v) Xenon trioxide acts as a strong oxidant.

$$XeO_3 + 6KI + 6HCl \longrightarrow Xe + 3I_2 + 6KCl + 3H_2O$$

(vi) Under the action of stronger oxidants, xenon trioxide gets changed to perxenate.

$$XeO_3 + 4NaOH + O \longrightarrow Na_4XeO_64 + O_2 + 2H_2O$$

Under normal conditions, it slowly gets decomposed into Xe, XeO_4, and oxygen. Xenic acid is a strong oxidizing agent as compared to ozone. The only by–product of its action is xenon gas, thus no extra complicating chemical species is introduced in the system when it is used as an oxidizing agent. The same is applicable for xenates and perxenates.It is an endothermic compound and has a large positive enthalpy of formation, $\Delta H_f = +401.7$ kJ mol^{-1} (c.f. DHf for $XeF_2 = -125.5$, $XeF_4 = -410$ and $XeF_6 = -401.7$ kJ mol^{-1}). The energy of Xe–O bond in xenon trioxide is approx. 84kJ mol^{-1}.

Xenon tetraoxide : It is an unstable gas at room temperature. It is obtained by the action of anhydrous sulphuric acid on Na_4XeO_6 or Ba_2XeO_6. At low temperature it occurs in the form of yellow solid. By its infrared examination, the XeO_4 was found to be tetrahedral.

$$Na_4XeO_6 + H_2SO_4 \longrightarrow XeO_4 + 2Na_2SO_4 + 2H_2O$$

$$Ba_2XeO_6 + 2H_2SO_4 \longrightarrow XeO_4 + 2BaSO_4 + 2H_2O$$

Perxenates of Group I and Group II Metals

Several perxenates of Group I and II metals have been isolated, and the crystal structures of $Na_4XeO_6.H_2O$ and $Na_4XeO_6.8H_2O$ have been determined by X–ray crystallography.

Sodium Perxenate, $Na_4XeF_6.8H_2O$: It has been prepared by the action of sodium hydroxide on xenon hexafluoride.

$$XeF_6 + 10NaOH \longrightarrow Na_4XeO_6 + 6NaF + 4H_2O + H_2$$

It is a white crystalline solid.

A large number of other salts of xenon are reported which do not find enough space in this small chapter.

The solubility of sodium perxenate in 0.5M NaOH is only 0.2 grams per litre, so precipitation of sodium perxenate could be used as a gravimetric method of analysis for sodium. Perxenates are extremely powerful oxidizing agents, which will oxidise HCl to Cl_2,H_2O to O_2, and Mn^{2+} to MnO_4. With concentrated H_2SO_4 they give xenon tetraoxide XeO_4, which is volatile and explosive.

Xenon Fluoride Complexes

XeF_2 acts as a fluoride donor and forms complexes with covalent pentafluorides including PF_5 AsF_5, SbF_5 and the transitions metal fluorides NbF_5, TaF_5, RuF_5, OsF_5, RhF_5, IrF_5 and PtF_5. These are thought to have the structure

$$XeF_2 \cdot MF_5 \quad [XeF]^+ [MF_6]^-$$

$$XeF_2 \cdot 2MF_5 \quad [XeF]^+[M_2F_{11}]^-$$

and $$2XeF_2 \cdot \quad MF_5 \quad [Xe_2F_3]^+[MF_6]^-$$

The structures of some of the XeF_2 complexes in the solid state are known. In the complex $XeF_2.2SbF_5$ the two Xe–F distances differ greatly (1.84 Å and 2.35 Å). This suggests the formulation $[XeF]^+[Sb_2F_{11}]^-$. However, the Xe–F distance of 2.35 Å is much less than the van der Waal's (non–bonded) distance of 3.50 Å. This suggests that one fluorine atom forms a fluorine bridge between Xe and Sb. In fact the structure is intermediate between that expected for the ionic structure and that for the fully covalent bridge structure.

XeF_4 forms only a few complexes, for example those with PF_5 AsF_5 and SbF_5.XeF_6 can act as a fluoride donor, forming complexes such as:

$$XeF_6 \cdot BF_3$$

$$XeF_6 \cdot GeF_4$$

$$XeF_6 \cdot 2GeF_4$$

$$XeF_6 \cdot 4SnF_4$$

$$XeF_6 \cdot AsF_5$$

$$XeF_6 \cdot SbF_5$$

XeF_6 may also act as a fluoride acceptor. With RbF and CsF it reacts as follows:

$$XeF_6 + RbF \longrightarrow Rb^+[XeF_7]^-$$

On heating, the $[XeF_7]^-$ ion decomposes:

$$2Cs^+ [XeF_7]^- \xrightarrow{50°C} XeF_6 + Cs_2[XeF_8]$$

Compounds of Krypton

At present the only fully characterised binary fluoride of krypton is the difluoride whereas there is some doubt about the existence of KrF_4.

When an electric discharge was passed through a mixture of krypton and fluorine at liquid temp, the product was reported to have the formation of KrF_4.

Repetition of this work was failed to confirm the existence of a tetrafluoride and has shown that under the condition of this experiment only the difluoride is formed.

(a) Krypton Difluoride : It is prepared by subjecting a mixture of krypton and fluorine in the ratio 1 : 2 by volume to a high voltage discharge at 7 – 12 mm Hg in a pyrex container at about – 180°C. The product of the reaction was colourless crystalline material

$$Kr + F_2 \longrightarrow KrF_2$$

It has been recently reported that KrF_2 can be synthesised photochemically if a mixture of krypton and fluorine gas is exposed to sunlight in pyrex glass bulbs.

Properties

(i) It is a colourless crystalline solid. It begins to sublime at about –60°C.

(ii) It is soluble in anhydrous HF at room temperatures.

(iii) It slowly decomposes to form the elements at room temperature but it can be stored for a long time in pyrex containers at solid CO_2 temperature.

$$KrF_2 \longrightarrow Kr + F_2$$

(iv) Unlike the difluoride of xenon, the hydrolysis of krypton difluoride in both aqueous base and acid is equally rapid.

$$KrF_2 + H_2O \longrightarrow Kr + \frac{1}{2}O_2 + 2HF$$

(v) Like XeF_2, KrF_2 dissolves in SbF_5 to give a yellow solution from which a white substance of the composition $KrF_2.2SbF_5$ has been isolated.

$$KrF_2 + 2SbF_5 \longrightarrow KrF_2.2SbF_5$$

When $KrF_2.2SbF_5$ is heated at 50°C, it yields krypton, SbF_5 and fluorine.

$$KrF_2.2SbF_5 \longrightarrow Kr + F_2 + 2SbF_5$$

There is enough evidence that AsF_5 combines with krypton difluoride. As KrF_2 can lose fluorine very easily at room temperature, it may therefore be used as a powerful fluorinating agent.

As it gets decomposed to yield fluorine at room temperature, it finds use as a fluorınating agent.

(b) Krypton tetrafluoride : There is doubt about the existence of krypton tetrafluoride.

However, it is prepared by passing electric discharge through krypton–fluorine mixture at low temperature.

$$Kr + 2F_2 \xrightarrow[\text{disch arge}]{\text{electric}} KrF_4$$

Krypton tetrafluoride is a colourless liquid. On hydrolysis at room temperature, it yields krypton, oxygen and hydrogen fluoride. However, on hydrolysis with ice at – 30°C it yields an acid of the composition $KrO_3.(H_2O)_x$ viz., H_2KrO_4, H_4KrO_5 or H_6KrO_6. Generally, the krypton compounds are less stable as compared to the corresponding xenon compounds, because of the greater ionization potential of the smaller xenon atom.

Compounds of Radon

Ionization potential of radon is even lower than that of Xe, so it also forms similar compounds. A radon fluoride has been prepared by heating the two elements. But as the half-life period of the most stable isotope of radioactive radon is 3 – 8 days, the composition and structure of radon fluoride have not been established. Theoretically oxides and chlorides of radon can also be synthesised. However, due to X–ray emission, radon compounds are too unstable to be studied.

1. Are Compounds of Helium and Neon Possible? Helium and neon are not able to form any compound. This can be explained in the following two ways.

(*i*) *In terms of V.B. theory.* According to V.B. theory, the formation of compounds of xenon can be explained on the basis of the excitation (promotion) of one, two or three of its 5*p* electrons to 5*d* orbitals. The resulting 5*p* and 5*d* unpaired electrons would then form bonds with fluorine or oxygen. As helium ($1s^2$) and neon ($1s^2 2s^2 p^6$) are not having vacant *d* orbital, these are unlikely to form any such compound.

(ii) *In terms of ionisation energies* as helium and neon are having high ionisation energies than the Ar, Kr, Xe and Rn, *helium and neon are having very little possibility for the formation of compounds.*

STRUCTURES OF INERT GAS COMPOUNDS

The nature of bonds and the orbitals involved in bonding in these compounds are of great theoretical interest. For xenon di fluoride structure and other compounds, the following four different possibilities are given as follows:

(i) *Xenon difluoride might be ionic in nature,* $^{-}F\overset{2+}{Xe}F^{-}$ *in which the outer shell of xenon is having six electrons.*

(ii) *The structure of xenon fluoride may be involving one electron bond of the following type.*

$$F-\overset{+}{Xe}F^{-} \longleftrightarrow {}^{-}F-\overset{+}{Xe}-F$$

The outer shell of xenon in this case will be having eight electrons.

(iii) *The structure of xenon fluoride may be involving two covalent bonds (valence bond theory).*

(iv) *Molecular orbital theory.*

Among the above four approaches, the fatter two have been used to explain the type of bonding in the inert gas compounds.

STRUCTURE OF XEF_2

Valence Bond (V.B.) Theory

The electronic configuration of xenon reveals that under normal conditions it fails to form any chemical compound. However, under excited conditions it can form three series of compounds which depend upon the stage of excitation. We will now explain this with the well-known fluoride of xenon, viz.

At. No. = 54, $1s^2$, $2s^2p^6$, $3s^2p^6d^{10}$, $4s^2p^6d^{10}$, $5s^2p^6$

The atomic number of xenon is 54. The electronic configuration of its outermost shell is $5s^25p^6$. When fluorine atoms come in contact with xenon, one of the electrons in $5p^6$ becomes unpaired and gets excited to 5*d* orbital. Thus, in xenon two unpaired electrons are available which form covalent bonds with two fluorine atoms to give rise to formation of XeF_2. In this compound there are two valence electrons, two in 5*s*, six in 5*p* and two in 5*d*. It means that in all five orbitals are required for filling these 10 electrons which can be formed by sp^3d hybridization. As XeF_2 involves sp^3d hybridization, it means that the structure of XeF_2 may be trigonal bipyramid with three equatorial positions occupied by the lone pairs. In order to confer the extra, stability on the compound, the shared electron pairs lie as far apart as possible. Xe – F bonds are 180° to each other and the compound assumes a linear shape, *i.e.,* F–Xe--F, with Xe–F distances 2.00 Å. The infrared and Raman spectra of XeF_2 show that the linear molecular structure is retained in the vapour.

Objections to Valence Bond Structure of XeF_2

(i) This explains the observed structure, but an objection is that the 5d orbitals of Xe appear to be too large for effective overlap of orbitals. The maximum in the radial electron distribution function for a 5*d* orbital in a Xe atom occurs at a distance of 4.9Å from the nucleus. It is known that the extent of *d* orbital participation in molecular bonding' that highly electronegative atoms like fluonrine cause a large contraction in the size of d orbitals. If this contraction is big enough, the valence bond explanation will suffice.

(ii) A second objection is over the mixing of orbitals (sp^3d hybridization). Mixing is only effective between orbitals of similar energy, and the Xe 5*d* orbitals would seem too high in energy to contribute to such a scheme of hybridization. (The difference in energy between a 5*p* and a 5*d* level is about 960 kJ mol^{-1}.)

CLATHRATE COMPOUNDS

Historical Introduction

The first known *clathrate,* discovered in 1811, was a gas hydrate of chlorine, which *Sir Humphrey Davy* believed to be a true chemical compound. Organic clathrate compounds have been known since 1849 and by 1900, a number of these species had been prepared. The nature of these materials was not understood until the X–ray work of Powell and his co–workers provided detailed structural information on a variety of examples. However, Powell defined the clathrate compounds as given below:

"Noble gases form compounds in which these gases are trapped within the cavities of crystal lattices of some inorganic and organic compounds. The compounds so formed are called clathrate compounds."

Powell showed that these compounds consisted of atoms or molecules of one substance called the *guest'* held in crystalline cages of another substance called the *'host'*. The 'guest' molecules are prevented from leaving the host cages by repulsive forces acting between the electron clouds of molecules forming the cavities, *i.e.,* '*host*' and of *'guest'* molecules. *The* '*guest*' can only escape from the 'host' when the forces holding the molecular cages together are overcome.

Why Name Clathrate?

Powell named these substance '*clathrates*' from the Latin word '*clathrates'* which means enclosed or protected by crossbars grating.

Modem Definition

The definition given by Powell is not general one and limited to inert gases only. However, it is observed that many clathrate compounds are known in which inert gases are not present. In its narrowest definition. "A clathrate is a single phase solid in which atoms or molecules of a guest component are completely surrounded and kept as in a cage by the crystalline lattice of host component without any sensible interaction between them except van der Wools forces." Many authors have broadened the above definition to include compounds in which the guest component is held as in an open channel or even as in a sandwich by the crystalline lattice of the host component. The crystalline lattice of the host component is generally much more expanded in its clathrate from that in pure form.

Stability of Clathrate Compounds

(i) The stability of many clathrate compounds is due to the formation, in the expanded lattice of the host itself, of strong interactions such as hydrogen bonds which do not exist to such an extent in the normal crystalline lattice of the host alone.

(ii) van der Waals forces and even sometimes weak charge transfer interactions contribute to the stability of clathrates.

(iii) Steric factors are always of prime importance because the clathrated molecule must adopt to the size and the shape of the rigid cage formed by the crystalline lattice of the host.

Characteristics of Clathrate Compounds

(i) These compounds are decomposed on melting or on dissolution of the compound in any suitable solvent.

(ii) Clathrate compounds cannot be considered to be compounds, since there are no specific chemical bonds between the *'guest'* and the '*host*' molecules, but there are only weak intermolecular forces.

(iii) According to X–ray studies, the lattice structure of the host in the clathrate is not in its normal crystalline form.

(iv) The clathrate form of the host lattice when its cavities are not occupied by guest atoms or molecules is usually less stable than the normal form.

(v) A clathrate compound is only formed when the guest molecules have such size and shape to fit the geometries of the cavities of host.

(vi) Although, the clathrates are not stoichiometric compounds, yet they can have a maximum composition formula.

$$nH.mG$$

In which m is the maximum number of guest atoms (G) or molecules which can be accommodated in a π cage of host molecule (H).

Examples of Clathrate Compounds

A few clathrates that have historical, theoretical or practical importance will be used to illustrate the subject.

1. Quinol–Noble Gases clathrates

Quinol (co–dihydroxybenzene) when crystallises from its solution in water or alcohol under a pressure of 20 atmospheres of a noble gas (argon, krypton or xenon) forms crystals of quinol-argon, quinol– krypton or quinol–xenon clathrate compounds. These quinol gas clathrate compound crystals differ form ordinary quinol crystals in their form and size. The noble gas in these quinol noble gas clathrates compounds is enclosed or entrapped in the crystal lattice of the quinol. *The properties of quinol- noble gas clathrates are as follows:*

(i) The crystals of these clathrates are quite stable. They can remain as such for several years.

(ii) On heating or dissolving in water, the noble gas is expelled leaving behind quinol as such. X–ray studies reveal that in crystal lattice of quinols, three molecules form a special cage of diameter 4 Å (4 × 10^{-8}cm).

In the quinol crystal, there are many such cavities connected together through very fine apertures within the crystal. Quinol, when crystallised from its aqueous or alcoholic solution in presence of other compounds, the atoms or molecules of which can fit tightly in its cavities form clathrate compounds. The entrapped molecules cannot come out.

(iii) The ratio of cavities to quinol molecules in the clathrate structure is 1:3 and the maximum composition formula expected is $[C_6H_4(OH)_2]_3Ar$. A total filling of the cages (I argon for 3 quinol molecules) is never reached during the crystallisation of quinol.

Structure of Quinol–Noble Gas Clathrates

Powell in England (1948) resolved the structure of these clathrates by X ray diffraction method. In these clathrates all the hydroxyl groups are tied together through hydrogen bridges, forming hexagons with oxygen atoms at each angle and the benzene rings pointing alternatively up and down, the interlocking of all hydroquinone molecules form an infinite three-dimensional cage work, the voids of which have a diameter of 7.6 A$^{\circ}$, big enough to entrap argon atoms.

In 1957 van der Waals predicted a neon/hydroquinone clathrate which might be possibly formed at pressure in excess of 160 atm. The first evidence for such a species was obtained by Gnauk (1964). The clathrate was obtained from a sorbitol/quinol solution and contained 4 wt percent of neon. Gnauk showed that all quinol clathrates crystallised from sorbitol contain higher noble gas concentrations than those crystallised from aqueous solutions. The structures also differ.

Helium does not form quinol clathrate. The atoms or molecules which are to be entrapped must be of suitable size (diameter of cavities in quinol's crystal lattice is approximately 7.9Å). The molecules should neither be too large (as they will not fit in) nor too small (as they will easily escape out). Helium atoms are believed to be too small and too difficult to deform or polarise (as required for van der Waals bonding) to be imprisoned in quinol cage.

2. Hydrates

(a) General Information: Villard and de Forcard (1925) prepared the first hydrates of argon, krypton, and xenon. Broadly speaking the cavities where the guest molecules are held in hydrates can be regarded as spherically symmetrical. There are tow structures I and II which are found in gas hydrates. In both structures the pentagonal dodecahedron is an essential unit. The dodecahedron exhibits degrees of distortion from their ideally regular forms.

In both structure the pentagonal dodecahedron is an essential unit. The dodecahedra exhibit degrees of distortion from their ideally regular form. Structure I contains 46 water molecules in the unit cell; there are two small cavities and six larger cavities. The small cavities are bounded by twenty water molecules while the larger cavities are enclosed by 24 water molecules. The small and large cavities have free diameters of about 5.1 and 5.8 A° respectively and each cavity is capable of holding one guest. This gives rise to the limiting theoretical composition as $5.75H_2O$. IG.

Structure II hydrates contain one hundred thirty six water molecules in the unit cell and there are sixteen small dodecahedral and eight larger hexadecahedral cavities. Their free diameters are 5.0 and 6.7 Å respectively. The large cavities often permit the formation of this structure type when the guest molecules are too large for the cavities in structure I when the hexadecahedral cavities are singly occupied, the limiting composition is $1.7H_2O$.IG.

(b) Types of Hydrates

(*i*) *Simple Hydrates.* Some simple noble gas hydrates are formed if water is crystallised slowly in the presence of the gases. No helium or neon hydrates have been prepared, even under high pressure, presumably because the helium and neon atoms are too small and too volatile to be retained in the cavities. Argon, krypton and xenon, however, form single-phase crystals of a structure I type in which n = 46 and m = 8 in the maximum composition formula $nH_2O.mNg$ where Ng is the noble gas.

Experimentally a ratio of $6H_2O$; 1 Ng is found which is very close to the theoretical value.

When water is allowed to freeze in the presence of noble gas (argon, krypton and xenon) under pressure, crystalline, hydrates are formed. The number of water molecules per noble gas is nearly 6. These noble gas hydrates are stable at very low temperatures. At higher temperatures, the noble gas escapes out. In these hydrates the covalently bonded water molecules are held together by hydrogen bonds. There is relatively little interaction between the noble-gas guest atoms and the water molecules cages.

(c) Helium and Neon do not Form Hydrates

(i) Helium and neon do not form hydrates, even under high pressure. Non–formation of helium and neon hydrates may probably be because of the fact that *helium and neon atoms are too small and too volatile to be retained in cavities.*

(ii) *Hydrates with heavy water.* Hydrates of krypton and xenon with heavy water have also been prepared.

(iii) *Double Hydrates.* Double hydrates are formed in the presence of small atoms (such noble gases) or molecules (such as H_2O, O_2, N_2, CO_2) then double hydrates are formed. These contain water, a large molecule (organic compounds like acetone, chloroform, carbon tetrachloride etc.) and a small molecule (noble gases). The general formula of each is:

$$\text{org. } 2G.\ 17H_2O$$

Org. = Molecule of acetone, chloroform, carbon tetrachloride or methylene chloride.

G = Noble gas (Ar, Kr, Xe).

In fact, the most remarkable aspect of these materials is their high decomposition at 1 atm. pressure. In each case the xenon containing double hydrate is the most stable because on account of the volatility of xenon occupies a greater proportion of the cavities and thereby has a greater stablising influence.

3. Radioactive Clathrates

Chleck and Ziegler (1959) prepared the krypton/hydroquionone clathrate from krypton containing 5 wt per cent of radioactive ^{85}Kr. Their final product had an activity of about three curies per gm. This was the first ^{85}Kr clathrate prepared. It is now known that krypton–85 can be stably incorporated into various solids and these are called *kryptonates. The* release of ^{85}Kr from the solid proceeds in proportion to the rate of any reaction occurring at the surface of the solid. This effect has been used to study reactions at the surface of a solid, such as corrosion rates, and for the analytical estimation of reactive gases.

The greatest application of kryptonates appears to have occurred in the realms of analytical chemistry. A quinol/Kr^{85} clathrate was prepared by heating solid quional to 186°C under a pressure of ^{85}Kr and allowing the mixture to cool slowly. The decomposition of this clathrate by ozone and chlorine dioxide was studied. Analysers suitable for the determination of ozone and atomic oxygen in the atmosphere have been devised by Chleck (1962). These system employ the hydroquinone/Kr^{85} clathrates.

4. Phenol Clathrates

Williams (1959) prepared the phenol clathrates of argon, krypton and xenon by direct combination of the gas with crystalline phenol and in the case of krypton and xenon with molten phenol. These results indicated that the ratio of argon and krypton to phenol may be 1 : 4 rather than 1 : 3 as predicted by Von Stackelberg (1958). The work done by Williams has produced a number of anomalies which still remain unresolved.

5. Complex Clathrnates

Three principal types of nickel Werner complexes form clathrates, but only with aromatic molecules. This fact suggests that van der Waals forces are not the only one to contribute to the stability of such clathrates but in some cases interactions of the charge transfer type between the host and guest molecules have been clearly demostrated. In such compounds, the clathrate character remains the most important because for molecules having the same charge transfer potential, it is their shape which is decisive for clathrate formation.

Now we will discuss the three types of such complexes;

(i) *Cyano momoammine nickel II clathrates.* The general formula of such clathrates of $Ni(CN)_2NH_3.M$ where M can be benzene, thiophene, furan, pyrrole, aniline or phenol. In its clathrate form this complex crystallises in planar arrays of nickel cyanide: the ammonia molecules, bonded to nickel atoms, project above and below each plane, delineating cavities between them; molecules larger than that mentioned above cannot be accommodated in the cages offering a purification process for the latter.

An ammoniacal solution of nickel cyanide productes a pale coloured clathrate when shaken vigorously with benzene. The formula $Ni(CN)_2.NH_3.C_6H_6$. The compound $Ni(CN)_2.NH_3.C_6H_6$ was originally thought to be co-ordination compound in which there were four groups bonded to the nickel, including the benzene molecule. It has been shown that the compound has a very interesting structure in which half of nickel ions have a co-ordination number of six because ammonia molecules are co-ordinated above and below the plane. The six co-ordinate nickel ions are bonded to the nitrogen atom of the CN^- group. The benzene molecule is trapped in the 'holes' of the lattice because it just fits in (see Fig. 4.18).

(ii) *Thiocyanate tetrakis* (4-methyl *pyridine*) *nickel* (*II*) *clathrates. Schaffer* (1957) has described many clathrates of this complex with benezene derivatives; they show a striking selectivity for para-isomers; pure meta xylene has been prepared on a semicommercial scale from a meta/para eutectic mixture by clathration of the para isomer. This complex forms a mole per mole clathrate by simple contact with para-xylene. This fact indicates that this clathrate could be of the channel type, although its structure is not yet known.

(ii) *Thiocyanato tetrakis* (*alpha–phenyl alkylamine nickel* (*II*) *clathrates*). *Hanotier* (1965) has prepared more than seventy complexes of the type $[Ni(CNCS)_2(y - C_6H_4)CH(R)(NH_2)_4]$ in which R is a primary alkyl group and y can be a wide variety of substituents on any position of the benzene ring; it is almost always possible to select one of those complexes for clathrating selectively any desired isomer in a mixture of aromatic molecules. These clathrates have been used to bring about the separation of optically active amines.

6. Other Clathrates

(a) Several compounds have a monocylic structure having in its centre a hole which can accommodate foreign molecules, *e.g.*, tri–o–thymotide, cyclodextrins, desoxycholic acid, cycloveratryl and few others. Graphite gives sandwich type inclusion compounds with alkali metals and with fluorine.

(b) Jeffrey and Mak (1965) have described an interesting intermediate between true clathrates and hydrogen–bonded hydrates. In hexamethylene tetramine, hexahydrate, $(CH_2)_6N_4 6H_2O$, the hexamethylene tetramine molecule is trapped inside a cage composed of water molecules.

(c) Powell and Welters (1959) prepared a clathrate of composition $(C_{18}H_{20}O_2)_6$ Ar, which on heating liberates more than fifteen times its own volume of gas (STP). The host substance is condensation product of phenol and mesityl oxide, called 4– (p–hydroxyphenyl) 2, 2, 4– trimethylchroman.

(d) A similar clathrate $(C_7H_8O)_6$ Xe was prepared by *Trofimov* and *Kazankin* (1965) by subjecting *p*– cresol to 10 – 50 atm. pressure of xenon. The heats of dissociation of the clathrate into liquid p–cresol and gaseous xenon and into solid p–cresol and gaseous xenon are 96.6 and 35.6 kJ respectively.

APPLICATIONS OF CLATHRATES

1. Clathrates of Werner complexes have been used to bring about the resolution of optically active amines.
2. Several factories, especially in U.S.S.R, use clathrates of urea for dewaxing petroleum products and for producing straightchain paraffins.
3. Several processes based on the formation of hydrates mostly propane–hydrate clathrate have been used for the desalinization of sea water.
4. The radioactive Kr^{85}/quinol has been proposed as a safe and convenient source of beta radiation.
5. Radiometric determination of gaseous fluorine in concentration of only a few parts per million has been made possible by the reaction of the fluorine with quinol/Kr^{85}clathrate.
6. Noble gas clathrates are used for storing noble gases. This eliminates both pressure and volume problems.
7. Clathrate compounds can be utilized to study the physical properties of isolated molecules under conditions when measurements could not normally be made.
8. Clathrate compounds have become important in the separation of hydrocarbons.

9. It is expected in the near future that noble gas clathrates might be used in handling and disposal of radioactive noble gases and as compact energy sources.
10. As helium and neon fail txo form clathrate compounds with hydroquinone whereas the Ar, Kr and Xe form such compounds, the former two inert gases can be separated from the latter three.
11. As the rare gas is immediately recoverable, either by heating or dissolution, the clathrates are used as temporary but convenient storage media in certain flow process.
12. Clathrate formation is of great use because it trapes the Kr^{85}, the radioactive isotope, produced by the fission of uranium in the nuclear reactors.
13. Kr^{85} having its long half–life of 10.3 years is found to be a convenient and compact source of energy for a small atomic battery, when used in the clathrate form.
14. The clathrate compounds can provide a very better way of examining megnetic susceptibilities of free molecules below the temperature at which they will normally condense to a liquid.

TYPES OF SILICATES

Silicates may be divided roughly into two types :

Soluble Silicates : These are derived only from the alkali metals and have composition commonly expressed in terms of varying amounts of metal oxide, silica and usually water.

Examples of soluble silicates are the sodium compounds, obtained generally by fusion in various proportion of silica with sodium carbonate at around 1000°C. Products obtained by fusion contain a variety of crystalline compounds like Na_2SiO_3, $Na_2Si_2O_5$, Na_4SiO_4, $Na_2Si_4O_7$, etc.

Soluble silicates are used in liquid detergent preparations to keep the pH high, so that grease and fat can be dissolved by forming a soap. Soluble silicates must not be used if the water is hard, or they will react with Ca^{2+} to form insoluble calcium silicate.

Sodium silicate is also used as an adhesive (for example for pasting paper, bonding paper pulp and corrugated cardboard), in asbestos roof tiles, in fireproof paint and putty, and in making silica gel.

5

Carbides and Nitrides

INTRODUCTION

The binary compounds of carbon with other elements are called carbides. However, this term has been employed mainly to include the binary compounds of carbon with metals and with only those nonmetals which form solids and refractory substances. Generally, non-metals form carbides which are either gases or low boiling liquids. Thus, this term excludes compounds with more electronegative elements such as phosphorus, oxygen, sulphur and halogen. Thus,

"Carbides are binary compounds in which carbon has combined with the elements of lower or about equal electronegativity."

Or

"Carbides are binary compounds in which carbon has combined with elements which are more electropositive than carbon."

TYPES OF CARBIDES

On the basis of the linkage involved, the carbides are classified into the following classes:

(A) Ionic or Electrovalent carbides.

(B) Covalent carbides.

(C) Metallic or Interstitial carbides. These are discussed one by one as follows

IONIC OR ELECTROVALENT OR SALT-LIKE CARBIDES

Ionic or salt like carbides are formed chiefly by the metals of groups IA, IIA, IIIA (except boron).

These are prepared by heating the metals or oxides or hydrides with carbon or with a source of ca bon such as carbon monoxide, carbon dioxide, an aliphatic or an aromatic hydrocarbon. These form transparent colourless crystals with ionic lattices containing metal cations in the interstices between carbon anions. These are easily decomposed by water or acids with the formation of the aliphatic hydrocarbons. Ionic or electrovalent carbides are further classified into the following types:

(i) **Acetylides :** These yield acetylene on hydrolysis and are believed to contain $[C{\equiv}C)^{2-}$ groups.

$$CaC_2 + 2H_2O \rightarrow Ca(OH)_2 + C_2H_2$$

This type of carbides may be regarded as derivatives of acetylene and are thus called acetylides. The alkali metals and copper, silver and gold form M_2C_2 type compounds. The alkaline earth metals, zinc and cadmium give MC_2 type carbides. The periodic group III metals form M_2C_6(M = Al) and MC_2 (M = Y, La, Ce, Pr, Nd) and a few miscellaneous compounds, *e.g.*, Th C_2, VC_2 and UC_2.

They possess face centred tetragonal space lattice with four molecules of RC_2 (R = Ca, Sr Ba) in the unit cell. This arrangement of atoms is present in a deformed lattice of sodium chloride or zinc sulphate. A slightly less symmetrical structure in an acetylide crystal is seen in that of thorium carbide.

(ii) **Methanides :** They yield methane on hydrolysis, *e.g.*, beryllium carbide.

$$Be_2C + 4H_2O \rightarrow 2Be(OH)_2 + CH_4 \uparrow$$

$$Al_4C_3 + 12H_2O \rightarrow 4Al(OH)_3 + 3CH_4 \uparrow$$

Methanides are regarded as the derivatives of methane and are said to contain C^{4+} groups. Beryllium carbide, Be_2 C (Be : : C : : Be) and aluminium carbide, Al_4C_3 are of this type

(iii) *Allylides :* They yield allylene on hydrolysis. The only example of this type is the magnesium carbide which contains C_3^{4-} discrete groups, *i.e.*, $(Mg^{2+})2$ (C = C = C$)^{4-}$

$$Mg_2C_3 + 4H_2O \rightarrow Mg\ (OH)_2 + C_3H_4$$

(iv) *Mixed carbides :* They yield a mixture of hydrocarbons on hydrolysis. Carbides of iron group and U_2C_3 belong to this class. These carbides are related in structure to intermetallic compounds.

Genera Methods of Preparation

The following are the main methods

(i) *By heating metal with carbon* : Metals like lithium and silver directly combine with carbon when heated at high temperatures.

$$2Li + C \xrightarrow{200^\circ C} Li_2C_2$$

$$2Ag + 2C \rightarrow Ag_2C_2$$

A number of other metals like titanium, vanadium, tungsten, molybdenum and iron combine with carbon to from carbides at 2200° C.

$$3Fe + C \xrightarrow{2200^\circ C} Fe_3C$$

$$3Mn + C \xrightarrow{2200^\circ C} Mn_3C$$

(ii) By heating oxides of metals with coke. A large number of carbides, *e.g.,* of Ca, B, Ba, Si, Al, W, U, Mn, etc., may be obtained by this method.

$$CaO + 3C \rightarrow CaC_2 + CO\uparrow$$

$$2Al_2O_3 + 9C \rightarrow Al_4C_3 + 6CO\uparrow$$

$$SiO_2 + 3C \rightarrow SiC + 2CO\uparrow$$

$$U_3O_8 + 14C \rightarrow 3UC_2 + 8CO\uparrow$$

(iii) By heating hydroxides, nitrides, phosphates, etc with carbon. Carbides of magnesium, calcium, and aluminium may be obtained by this method at 2000°C.

$$2Mg(OH)_2 + 4C \rightarrow 2MgC_2 + 2H_2O + O_2\uparrow$$

$$Ca_3N_2 + 6C \rightarrow 3CaC_2 + N_2$$

$$4AlPO_4 + 19C \rightarrow Al_4C_3 + 4P + 16CO\uparrow$$

(iv) By passing acetylene or ethylene on heated metals or its oxide.

$$4Li + C_2H_2 \rightarrow Li_2C_2 + 2LiH$$

$$Ag_2O + C_2H_2 \rightarrow Ag_2C_2 + H_2O$$

(v) By passing acetylene gas into ammoniacal solution of salt of certain metals, *e.g.,* Cu, Ag, Au, etc.

$$2CuCl + 2NH_4OH + C_2H_2 \rightarrow Cu_2C_2 + 2NH_4Cl + 2H_2O$$

$$2AgNO_3 + 2NH_4OH + C_2H_2 \rightarrow Ag_2C_2 + 2NH_4NO_3 + 2H_2O$$

(vi) By passing carbon monoxide over the metals like aluminium etc. at high temperature.

$$6Al + 2CO \rightarrow Al_4C_3 + Al_2O_3$$

(vii) By heating metallic oxides or other salts with calcium carbide.

$$CaC_2 + MgO \rightarrow MgC_2 + CaO$$

$$CuCl_2 + CaC_2 \rightarrow CuC_2 + CaCl_2$$

(vii) By heating carbonate or oxide with carbon in the presence of very oxidisable metal.

$$BaCO_3 + 3Mg + C \rightarrow 3MgO + BaC_2$$

General Properties

(i) Alkali metal carbides are generaly soft and transparent when pure. These are easily decomposed by water or acids with the formation of aliphatic hydrocarbons. They are non-conductors of electric current in the solid state.

(ii) Coinage metal carbides are coloured *e.g.,* Cu_2C_2 (dare brown), Ag_2C_2 (white) and Au_2C_2 (yellow) and all are explosive powders.

(iii) Alkaline earth metal carbides get decomposed by dilute acids or water to yield acetylene gas. They combine with nitrogen when heated in it.

$$CaC_2 + 2H_2O \rightarrow Ca(OH)_2 + C_2H_2$$

$$CaC_2 + N_2 \rightarrow CaCN_2 + C$$

They are able to reduce magnesium oxide to magnesium.

$$3MgO + CaC_2 \rightarrow 3Mg + CaO + 2CO \uparrow$$

They react with ammonia and bromine as follows :

$$CaC_2 + 4NH_3 \rightarrow CaCN_2 + NH_4CN + 4H_2 \uparrow$$

$$CaC_2 + 4Br_2 \rightarrow CaBr_2 + C_2Br_6$$

Beryllium carbide is a brick red coloured compound. It is hard enough to scratch glass and quartz,

(iv) Aluminium carbide is pale yellow hard crystalline substance M.P. 2200°C. It is decomposed by water liberating methane. It reacts with nitrogen at 1800°C to form Al_5C_3N and with oxygen to form alumina.

$$Al4C_3 + 6O_2 \rightarrow 2Al_2O_3 + 3CO_2.$$

COVALENT CARBIDES

The only true covalent carbides are those of silicon, SiC and boron B_4C. These are not attacked by decomposed at very high temperatures. Because of their hardness they are used for cutting and as abrasives. Boron and silicon carbides may be prepared by reducing the oxides with carbon in an electric furnace.

METALLIC OR INTERSTITIAL CARBIDES

These are chemically inert and extremely hard like diamond. They possess metallic lustre, high electrical conductivity and are therefore called metallic carbides. Interstitial carbides are formed by the transition metals of the periodic groups IVB. VB and VIB. These are of two types :

MC (M = Ti, Zr, Hr, V, Nb, Ta, Mo, W) and M_2C
(M = V, Mo and W).

The metals. *e.g.,* Cr, Fe, Ni and Co possess smaller atomic radii and therefore do not form typical interstitial carbides. Although, the carbides of these elements are similar to the interstitial carbides yet they are distinguished from the latter by their reactions with water or dilute acids to give either simple hydrocarbons or mixture of methane, ethylene, ethane and even liquid and solid hydrocarbons and free carbon. These carbides are thus, intermediate between ionic and typical interstitial carbides. Their structure is roughly described as having carbon chains running through very distorted metal lattices.

Preparation

They are prepared by heating the powdered metal to a high temperature at about 2200°C with carbon.

Properties

(i) These carbides are extremely hard, some approaching to diamond in hardness and have high melting point.

(ii) They are having high electrical conductivity.

(iii) Many of them show extreme chemical inertness even under drastic conditions.

(iv) The carbides are most likely to get attacked by oxidising agents. Molybdenum and tungsten carbides are somewhat more reactive chemically than the metals from with they are derived, but the other carbides are somewhat less reactive than the parent metals.

Structure

Chemically the refractory carbides and solid phases rather than dividual molecular species with the atoms held toghter by chemical onds.

Nearly all metals are known to crystallize with their atoms in hexagonal cubic close-packing and interstitial carbides etc. can be formed by carbon atoms entering the octahedral holes. Hence the relative size of the carbon atoms and the metallic atoms constitute an important factor, the ratio of the radius of the metallic atom to that of the non-metallic must be 1:0.41 – 0. 59. For this to be possible without distortion of the lattice, the radius of the close packed metal atoms should be not less than 1.3Å, which is the case with al the metals which form MC type carbides. The atomic radii of iron. cobalt, chromium, manganese and nickel are about 1.25Å, so that the octahedral holes are too small to accommodate carbon atoms.

Hence, although iron has been reported to form, iron cementite, Fe_3C the structure of which does not resemble that or iron, and it is different from the other refractory carbides in its chemical activity. Nickel cementite, Ni_3C is much less stable and cobalt cementite, Co_3C cannot be prepared although the phase diagram of cobalt and carbon reveals its existence under certain conditions. It is evident that composition of these substances like those of the intermetallic compounds could not be determined by the valency of the metal and non-metal components in the ordinary sense. Hence it can be said that, a number of attempts to describe the bonding in these refractory materials in terms of more familiar concepts have been made but the problem must still be regarded an open one. At present it is regarded that the bonds in these carbides are having the partial character both of covalent bonds (viz. these materials are often as carborundum) and to metallic bonds (viz. substantial conductivities of the solids).

PREPARATION AND PROPERTIES OF VARIOUS CARBIDES

1. First Group Carbides

(a) *Alkali metal carbides :* These metals form electrovalent or ionic carbides

Lithium carbide can be prepared by heating lithium with carbon : sodium carbide is formed in this way but the reaction is reversible and the yield is poor; potassium, rubidium and caesium react with carbon to yield interstitial carbides like C_8K.

$$2Li + 2C \rightarrow Li_2C_2$$

$$2Na + 2C \rightarrow Na_2C_2.$$

$$8C + K \rightarrow C_8K$$

All the alkali metal carbides may be prepared by passing acetylene over heated metal at red heat.

$$2Li + C_2H_2 \rightarrow Li_2C_2 + H_2$$

$$2Na + C_2H_2 \rightarrow Na_2C_2 + H_2$$

Alkali metal carbides are usually soft and transparent when pure. These are non-conductors of electric current in the solid state. These are easily decomposed by water or acids yielding aliphatic hydrocarbons. These are strong reducing agents.

(b) *Coinage metal carbides* : These cannot be prepared by direct combination of the metals with acetylene but are produced by passing acetylene into an ammonicacal solution of the salt of the metals.

$$2Cu(NH_3)_2Cl + C_2H_2 \rightarrow 2NH_4Cl + Cu_2C_2 + 2NH_3$$

$$2Ag(NH_3)_2NO_3 + C_2H_2 \rightarrow 2NH_4NO_3 + Ag_2C_2 + 2NH_3$$

Copper acetylide is dark brown and explodes if heated to 120° or in contact with oxidising agents. It liberates acetylene when dissolved in dilute HCl or KCN solution. Silver acetylide is white and is more explosive than copper acetylide. Again, it reacts with nitric acid or sulphuric acid to regenerate acetylene.

Au_2C_2 is yellow and explosive.

2. Second Group Carbides

(a) Beryllium Carbides

Beryllium forms two carbides Be_2C and BeC_2.

(i) *Beryllium acetylide BeC_2* : It is prepared by passing acetylene over heated metallic beryllium at 450°C.

It reacts with water, alkalies or acids liberating acetylene.

$$Be + C_2H_2 \xrightarrow{450^\circ C} BeC_2 + H_2$$

(ii) Beryllium methanide Be_2C.

It is prepared by heating the elements together at 1300°C. It is a brick red compound, hard enough to scratch glass and quartz. It is decomposed by water, dilute acids and alkalies liberating methane.

$$2Be + C \xrightarrow{1300^\circ C} Be_2C$$

$$Be_2C + 4H_2O \rightarrow 2Be(OH)_2 + CH_4 \uparrow$$

$$Be_2C + 4HCl \rightarrow 2BeCl_2 + CH_4 \uparrow$$

$$Be_2C + 4HCl \xrightarrow{600^\circ C} 2BeCl_2 + C + 2H_2 \uparrow$$

$$Be_2C + 4HF \xrightarrow{450^\circ C} 2BeF_2 + C + 2H_2 \uparrow$$

Structure : The lattice of beryllium is antifluorite.

(b) Magnesium Carbides

Magnesium forms two types of carbides, MgC_2 and Mg_2C_3.

(i) Magnesium Acetylide MgC_2

Magnesium acetylide may be prepared by heating magnesium in acetylene at 500°C or by the action of acetylene on an ethereal solution of magnesium diethyl.

$$Mg + C_2H_2 \xrightarrow{500^\circ C} MgC_2 + H_2$$

It is also prepared by heating MgO and $Mg(OH)_2$ with carbon or calcium carbide.

$$MgO + 3C \xrightarrow{2000^\circ C} MgC_2 + CO$$

$$Mg(OH)_2 + 4C \xrightarrow{2000^\circ C} MgC_2 + 2CO + H_2$$

$$MgO + CaC_2 \rightarrow MgC_2 + CaO$$

It is having steel grey colour and is hydrolysed liberating acetylene gas.

$$MgC_2 + 3H_2O \rightarrow C_2H_2 + Mg(OH)_2$$

It is unstable at high temperatures.

$$2MgC_2 \rightarrow Mg_2C_3 + C$$

Structure. It is a tetragonal crystalline solid.

(ii) Magnesium carbide Mg_2C_3

It is obtained by the action of methane or propane on magnesium at 700°C or by the action of acetylene on magnesium at 600°C.

$$2Mg + 3CH_4 \xrightarrow{700^\circ C} Mg_2C_3 + 6H_2$$

$$4Mg + 3C_2H_2 \xrightarrow{600^\circ C} 2Mg_2C_3 + 3H_2$$

It is readily hydrolysed by water yielding allkylene.

$$Mg_2C_3 + 4H_2O \rightarrow 2Mg(CH)_2 + CH_3 - C \equiv CH$$

Structure. It is hexagonal crystalline solid.

(c) Acetylides of Calcium, Barium and Strontium

All these metals from carbides or acetylides os the type MC_2. These are prepared by heating the oxides with carbon in an electric furnace.

$$MO + 3C \xrightarrow{\text{Electric furnace}} MC_2 + CO$$

$$(M = Ca, Ba, Si)$$

$$BaO + 3C \rightarrow BaC_2 + CO$$

$$CaO + 3C \rightarrow CaC_2 + CO$$

The most important is calcium carbide which may also be obtained by the action of carbon on phosphate or nitribe of the metal at 2000°C.

$$Ca_3N_2 + 6C \xrightarrow{2000^\circ C} 3CaC_2 + N_2$$

$$Ca_3(PO_4)_2 + 14C \xrightarrow{2000^\circ C} 3CaC_2 + 2P + 8CO$$

Barium carbide may be prepared as follows :

$$BaSO_4 \xrightarrow[\text{heat}]{\text{red}} BaC_2 + SO_2 + O_2$$

$$BaCO_3 + 3Mg + C \rightarrow 3MgO + BaC_2$$

These carbides are white crystalline solids in pure state. They are decomposed by water or dilute acids to form acetylene.

$$CaC_2 + 2H_2O \rightarrow Ca(OH)_2 + C_2H_2$$

These are powerful reducing agents

$$3MnO + CaC_2 \rightarrow 3Mn + CaO + 2CO$$

Calcium acetylide or carbide is a strong reducing agent, *i.e.*, it reduces magnesium oxide to magnesium.

$$3MgO + CaC_2 \rightarrow 3Mg + CaCO + 2CO$$

When red heated, it absorbs nitrogen to form calcium cynamide.

$$CaC_2\ N_2 \rightarrow CaNCN + C$$

The acetylides of strontium and barium resemble those of calcium.

$$BaC_2 + N_2 \xrightarrow[800^\circ C]{700^\circ C} Ba(CN)_2$$

$$Ba(CN)_2 \rightarrow BaCN_2 + C$$

$$Ca(CN)_2 \rightarrow CaCN_2 + C$$

Structure. The acetylates of calcium, barium and strontium posses face-centred tetragonal space lattice with four molecules of RC_2 (Ca, Ba, Sr, R) in the unit cell. This arrangement of atoms is present in a deformed lattice of sodium chloride or zinc sulphate.

(d) Carbides of the Zinc Family

Zinc carbide. It is a white powder and is obtained by passing acetylene into a solution of zinc diethyl in ligroin (ZnC_2). It is also prepared by heating zine metal with acetylene at 400 – 500°C

$$ZnC_2 + 2H_2O \rightarrow Zn(OH)_2 + C_2H_2$$

It is a white powder and hydrolysed with water to form corresponding hydroxide and acetylene.

$$Nz + C_2H_2 \xrightarrow{400-500^\circ C} ZnC_2 + H_2$$

Cadmium Carbide

It is prepared by heating the metal with acetylene at 400-500°C. It is similar to zinc carbide.

Mercury Carbides

Carbides of mercury are very ill defined. However, two carbides of mercury are known.

(i) HgC_2. It is prepared by passing gas into an aqueous solution of mercuric cyanide. It is highly explosive.

(ii) Hg_2C_2. It is obtained by passing acetylene into mercurous acetate solution.

The properties and structures of mercury carbides require investigation.

3. Third Group Carbides

(i) Boron carbides B_4C or $B_{12}C_3$. It is a covalent carbide and is obtained by heating amorphous boron or boric anhydride with carbon in electric furnace of 3000°C.

$$4B + C \rightarrow B4C$$

$$2B_2O_3 + 7C \rightarrow B_4C + 6CO$$

(ii) It may also be prepared by dissolving carbon and boron in certain metals like copper or silver in an electric furnace. The mass obtained is then treated with nitric acid to dissolve the metal.

(iii) It is prepared by passing mixture of hydrogen and boron trichloride through the glass globe at 3500°C. Boron carbide formed is deposited on the water cooled brass tubes. In the glass globe, there is carbon strip with brass tubes and water cooling arrangement.

Properties

(i) It is a black, very hard crystalline substance. It melts at 2350°C.

(ii) It is not attacked by chemical reactions at ordinary temperatures. However, it is attacked by oxygen and chlorine at 1000°C.

(iii) It reacts with molten potassium hydroxide forming carbon monoxide.

Structure

Boron carbide is a covalent carbide and has a peculiar structure. Clark and Horad (1943) established its structure. According to them,

(i) Boron carbide forms straight chains of three carbon atom and a group of twelve boron atoms which are being arranged at the vertices of a nearly regular icosahedron.

(ii) Each boron atom is bonded to five other boron atoms in the same icosahedryl group and is also bonded either to a carbon atom or to a boron in an adjacent icosahedron. Thus, a continuous three-dimensional network of boron atoms extends throughout the crystal.

(iii) The carbon atom at the end of carbon chain is attached to the central carbon atom and to three boron atoms; the central carbon atom is attached only to the other two carbon atoms of the chain. The bond distances are B – B = 1.74 to 1.80Å; B – C = 1.64AÅ and C – C = 1.39Å.

Uses : It is used :

(i) In cutting diamonds

(ii) For drilling holes in rocks

(iii) For making electrodes of electric furnaces.

(iv) For making lamp filaments.

(d) Carbides of Aluminium

It forms two carbides :

(i) Al_4C_3

Preparation

1. It is prepared by heating a mixture of carbon and aluminium at temperature above 1000°C. The product is extracted with the cold concentrated hydrochloric acid which dissolves unreacted aluminium. It is then dried over ether.

$$4Al + 3C \rightarrow Al_4C_3$$

2. It may also be prepared by heating aluminium oxide or aluminium phosphate with carbon.

$$2Al_2O_3 + 9C \rightarrow Al_4C_3 + 6CO$$

$$4AlPO_4 + 15C \rightarrow Al_4C_3 + 4P + 16CO$$

3. It is prepared by the action of carbon monoxide on aluminium at high temperature.

$$6Al + 3CO \rightarrow Al_2O_3 + Al_4C_3$$

4. It is prepared by action of acetylene on aluminium at high temperature.

$$2Al + 3C_2H_2 \xrightarrow{450-500^\circ C} Al_2C_6 + 3H_2$$

Propertics

Al_4C_3 forms yellow hexagonal crystals. It sublimes at 2200°C without decomposition and is hydrolysed by water on warming giving methane.

$$Al_4C_3 + 1lH_2O \rightarrow 4Al(OH)_3 + 3CH_4$$

With nitrogen at 1800°C, it forms Al_5C_3N.

It reacts with oxygen giving alumina.

$$Al_2C_3 + 6O_2 \rightarrow 2Al_2O_3 + 3CO_2$$

Structure

The crystals of aluminium carbide (Al_4C_3) are regular hexagons. In the space lattice of Al_4C_3 crystal, each aluminium atom is being surrounded by three carbon atoms and each carbon by four aluminium atoms. The distance of each carbon atom from aluminium is 1.90 to 2.22Å where a the shortest C–C distance is 3.16Å. Therefore, the carbon atoms are not linked directly to one another.

(ii) *Al_2C_6* : It is obtained by passing acetylene on the surface of aluminium metal at about 500°C.

$$2Al + 3C_2H_2 \xrightarrow{450-500^\circ C} Al_2C_6 + 3H_2$$

This compound could not be obtained in the pure condition. Al_2C_6 is slowly hydrolysed by water forming acetylene.

$$Al_2C_6 + 3H_2O \rightarrow Al_2O_3 + 3C_2H_2$$

Structure

Al_2C_6 possesses crystal structure similar to that of calcium carbide.

(c) *Rare earth carbides* : All the rare-earths form carbides of the type MC_2, where M is any rare earth metal. However, cerium also forms carbide of the type Ce_2C_6.

Rare earth carbides may be prepared by heating the mixture of metal or its oxide and carbon in an electric furnace.

$$CeO_2 + 4C \quad 2CO + CeC_2$$

$$Ce + 2C \rightarrow CeC_2$$

Cerium carbide Ce_2C_6 is prepared by heating cerium in a current of acetylene.

$$2Ce + 3C_2H_2 \rightarrow Ce_2C_6 + 3H_2$$

Carbides of thorium and uranium are prepared as follows :

$$ThO_2 + 4C \rightarrow ThC_2 + 2CO$$

$$U_3O_8 + 14C \rightarrow 3UC_2 + 8CO$$

These carbides are reactive. They yield acetylene gas on the reaction with water.

$$3MC_2 + 6H_2O \rightarrow 2M_2O_3 + 2C_2H_2 + CH_4 + > CH_2$$

The radical $> CH_2$ may polymerise either to olefines or to cyclic hydrocarbons. ThC_2 is yellow crystalline solid and decomposes in moist air. UC_2 is very hard, scratching glass and quartz. It oxidises at 370°C very rapidly to form uranium oxide and carbon dioxide.

$$3UC_2 + 10O_2 \xrightarrow{370^\circ C} U_3O_8 + 6CO_2$$

It emits sparks when struck and ignites on powdering quickly. It reacts with nitrogen to form uranium nitride. The structure of these carbides is similar to calcium carbide. The structure of ThC_2.

4. Carbides of Fourth Group

Elements of fourth group form a number of refractory and hard carbides.

(a) Carbon Carbides

Carbon also forms a carbide, *i.e.,* carbon carbide or diamond. It is very hard and is used for cutting, boring and for abrasive purposes.

(b) Silicon carbide SiC

It is also known as carborundum.

Preparation : It is prepared by heating a mixture of coke and silicon dioxide in an electric furnace at 2000°C.

$$SiO_2 + 2C \rightarrow Si + 2CO$$

$$Si + C \rightarrow SiC$$

$$SiO_2 + 3C \rightarrow SiC + 2CO$$

(ii) It may also be obtained by passing acetylene on heated silicon.

$$2Si + C_2H_2 \rightarrow 2SiC + H_2$$

(iii) It is also obtained by heating the mixture of iron silicide acid excess of carbon in an electric furnace.

Properties

(i) It is colourless when it is pure. The commercial sample is yellow, green or blue.

(ii) It does not undergo decomposition below 2200°C.

(iii) It is very hard. Its hardness is between that of the ruby and diamond.

(iv) It is not attacked by HF, HCl. etc.

(v) When fused with NaOH on contact with air, it gets oxidised.

$$SiC + 4NaOH + 2O_2 \rightarrow Na_2Sio_3 + Na_2CO_3 + 2H_2O$$

Uses : It is used as an abrasive and is also used for cutting.

Structure

Silicon carbide exists in three forms which are different combinations of 'layers' corresponding to the zinc blende and wurtzite structures. The forms are :

Carborundum	I	aaa	bb
	II	aaa	bbb
	III	aa	bb

where a is "zinc blende layer" and b is a "wurtzite layer".

The lattice of the silicon carbide consists of carbon atoms at points corresponding to those occupied by atoms in a close packed face-centred cubic or hexagonal structure with silicon atoms at half the points corresponding to the position of the tetrahedral holes. This type of structure of silicon carbide has been confirmed by X-ray studies.

(c) Carbides of Germanium, Tin and Lead

These elements possess little tendency to form carbides. However, lead forms a carbide of the formula PbC_2 which is prepared by the action of calcium carbide on a lead acetate solution.

$$Pb(CH_3COO)_2 + CaC_2 \rightarrow PbC_2 + Ca(CH_3COO)_2$$

Lead carbide is a green powder. It yield s acetylene on treatment with hydrochloric acid.

$$PbC_2 + 2HCl \rightarrow PbCl_2 + C_2H_2$$

(d) Carbides of Titanium, Zirconium and Hafnium

The carbides can be prepared by heating their oxides with carbon in an electric furnace at 1700 – 18000°C.

They are refractory carbides. $TiO_2 + 2C \rightarrow TiC + CO_2$

5. Fifth Group Carbides

(a) *Vanadium carbides :* Vanadium forms carbides V_2C, VC and VC_2 but the best known carbide is VC. It is prepared by heating vanadium or oxide with carbon in an electric furnace. It is an example of refractory carbides. It gets decomposed by water and acids at red heat. It is however slowly decomposed by cold nitric acid.

$$V + C \xrightarrow{2000^{\circ}C} VC$$

(b) *Tantalum carbide :* Tantalum carbide is extremely hard and is metallic in appearance. It is best prepared by heating tantalum with carbon at about 2200°C, Tantalum carbide is highly refractory and exhibits super conductivity. It is paramagnetic. It is regarded as an excellent substance for the construction of cutting tools, wire dice, etc.

$$Ta + C \xrightarrow{2200^\circ C} TaC$$

Its filaments are about 40% brighter at the melting point of the metal.

Niobium carbide : It is similar to vanadium carbide.

6. Sixth Group Carbides

The elements of VIB group (Cr, Mo & W) form different types of carbides as :

Cr_3C_2	Mo_2C	W_2C
Cr_4C_2	MoC	WC
Cr_5C_2	MoC_2	W_5C_2

(a) *Carbides of chromium :* Chromium forms carbides Cr_3C_2, Cr_4C_2 and Cr5O2 but the best known carbides is Cr_3C_2.

$$3Cr_2O_3 + 13C \rightarrow 2Cr_3C_2 + 9CO$$

It is refractory and hard. It is inert and not attacked by acids or alkalies. In the crystal of Cr_2C_3, each carbon atom is surrounded by six chromium atoms which lie at the corners of a trigonal biprism containing it, and a given carbon atom is only 1.665Å from two other carbon atoms. The arrangement of carbon atoms may be compared with a hydrocarbon chain which is stripped of H atoms.

(b) *Tungsten Carbides :* Tungsten forms three carbides :

(i) W_2C

(ii) WC

(III) W_5C_2

The best characterised carbide of tungsten is WC.

It may be prepared :

(i) By heating tungsten metal with gases containing carbon at 800°C or

(ii) By heating tungsten oxide with carbon in the presence of iron which reduces the reaction temperature or,

(iii) By heating tungstic acid or ammonium paratungstate in a mixture of methane and hydrogen at 850°C.

It is brittle and used for the preparation of carbides :

(c) *Uranium carbides* : It forms two carbides :

(a) *UC* : It is prepared by heating uranium which carbon at 2100°C. It has the sodium chloride structure.

(b) *UC_2* : It can be prepared by heating uranium with excess of carbon at 2300°C.

$$U + 2C \xrightarrow{2400^\circ C} UC_2$$

It may also be obtained by heating 8 parts of charcoal with 50 parts of U_3O_8 in an electric furnace.

It possesses metallic appearance. It ignites in air at 400°C. It decomposes with cold water liberating hydrocarbons. It is rapidly oxidised by air at 370°C.

It possesses a crystal similar to that of chloride. It is very hard.

U_2C_3 is also known.

7. Seventh Group Carbides

Only managanese forms carbide having the composition Mn_3C. It is obtained by heating manganese or its oxide with carbon at a high temperature.

$$3Mn + C \rightarrow Mn_3C + 54.2kJ$$

It is crystalline solid. It undergoes hydrolysis to form methane and hydrogen.

$$Mn_3C + H_2O \rightarrow 3Mn(OH)_2 + CH_4 + H_2$$

It also reacts with halogens, ammonia, nitric oxide, hydrochloric acid, etc.

8. Eight Group Carbides

Platinum metals do not form any definite type of the carbides but metals of iron family form carbides. For example,

(i) Iron Carbides Fe_3C : It is also known as cementite which is readily formed by dissolved carbon in molten iron.

$$3Fe + C \rightarrow Fe_3C$$

When the melt crystallises, Fe_3C separates out.

It forms crystals of metallic appearance.

Fe_3C possesses metallic lusture. Its specific gravity is 6.94 and melting points is –1300°C. It burns in oxygen at 150°C. It is decomposed by steam at 400-500°C. It reacts with water and dilute hydrochloric acid to give hydrocarbons and hydrogen.

With HCl acid, it gives a mixture of gaseous and liquid hydrocarbons along with hydrogen. It gets decomposed by water with the deposition of carbon and the formation of ethane, ethylene and other hydrocarbons.

(ii) Cobalt carbide Co_3C: This compound could not be obtained in the solid condition. However, its existence has been inferred from the thermal analysis of CO – C system.

(iii) Nickel Carbide : This compound could not be prepared so far.

INDUSTRIAL APPLICATIONS OF CARBIDES

1. The extreme hardness, very high melting point (usually between 3000 and 4000°C) and chemical inertness of carbides render the refractory carbides of considerable importance especially for making tools.
2. Silicon carbide being used as an abrasive and a deoxidant in the metallurgy and as resistors for electric furnaces.
3. Carbides have been used as reducing agent in metallurgy and in electro-thermal work.
4. Aluminium carbide is used for the preparation of methane whereas calcium carbide for acetylene.
5. As carbides ca withstand high temperatures, they are widely used in furnace lining.
6. Tungsten and tantalum carbides are used for the manufacture of high speed cutting tools.
7. Boron carbide is being used for -
   (i) Cutting diamonds,
   (ii) For drilling holes in rocks,
   (iii) For making electrodes of electric furnaces, and

(iv) For making lamp filaments.

NITRIDES

Introduction to Nitrides

The binary compounds of nitrogen with electropositive element or elements which are less electronegative than nitrogen are known as nitrides.

The above definition of nitrides does not include the salts of hydrazoic acid (*e.g.*, NaN_3) which are not nitrides. A large number of nitrides of both metals and non-metals are known. Their properties vary widely and depend on the type of bonding the metal and nitrogen atom.

CLASSIFICATION OF NITRIDES

The intrides have been classified into following types depending upon the nature of bonding between the metal and nitrogen atom :

(a) Ionic Nitrides

These nitrides are derived from trinegative nitrogen anion (N^{-3}) having anionic radius of 1.71Å and are characterised by the usual valency formulae. These nitrides are crystalline compounds having high melting points. Ionic nitrides are formed by lithium, copper (I) and second group metals except mercury and thorium. These are hydrolysed by water to yield hydroxide and ammonia.

$$Li_3N + 3H_2O \rightarrow 3LiOH + NH_3$$

$$Ca_3N_2 + 6H_2O \rightarrow 3Ca(OH)_2 + 2NH_3$$

Sodium and potassium nitrides are known but are difficult to prepare. The existence of mercury nitride is doubtful. Ionic nitrides form colourless crystals and have high melting points.

When heated in a current of carbon dioxide at 200°C, nitrogen gas is evolved.

$$Mg_3N_2 + 3CO_2 \rightarrow 3MgO + 3CO + N_2$$

When heated in a current of H_2S, sulphides are obtained.

$$Mg_3N_2 + 4H_2S \rightarrow 3MgS + (NH_4)_2 S$$

Ionic nitrides are prepared by the following methods :

(i) By the direct union of metal with nitrogen,

$$6Li + N_2 \rightarrow 2Li_3N$$

$$3Mg + N_2 \rightarrow Mg_3N_2$$

(ii) By passing ammonia over heated metal or its oxide.

$$3Mg + 2NH_3 \rightarrow Mg_3N_2 + 3H_2$$

$$3Ag_2O + 2NH_3 \rightarrow 2Ag_3N + 2H_2O$$

(iii) By deammonation of the amides are elevated temperatures.

$$3Ba(NH_2)_2 \xrightarrow{\text{Heat}} Ba_3N_2 + 4NH_3$$

A large number of double nitrides are known. Examples of these are: LiMgN, LiZnNi, Li_3AlN_3, Li_5SiN_2, Li_5TiN_3, Li_5GeN_3 and Li_3GaN_2. These are pale in colour, thermally stable but easily hydrolysed by dilute HCl acid. Double nitrides are obtained by heating lithium nitride with the nitride of another metal. They get crystallised with co-ordination lattice structures and are salt-like in character.

The formulae of the ionic nitrides correspond to what would result form combination of the normal metal ions with N^{-3} ions (radius 1.71Å), viz. Li_3N, CaN_2, Ba_3N_2, etc. The exact structure of these nitrides is somewhat doubtful, the structure of lithium nitride Li_3N is generally written a $Li^+[Li–N^-–Li]$.

Lithium nitride has a hexagonal structure with 1/3 of the lithium ions having two nitrogens at 1.94 Å and the other three at 2.11Å. The nitrogens have two lithium neighbours at 1.94Å and six at 2.11Å. Nitrides of the type M_3N_2 are anit-isomorphous with oxides of the type M_2O_3 metal atoms occupying oxygen positions and nitrogen atoms occupying metal positions in the oxide structure.

(b) Interstitial or Metallic Nitrides

These are the true interstitial compounds with nitrogen atoms occupying the interstices in the metal structures. These nitrides have atomic radii sufficiently great to allow the inclusion of nitrogen atoms in the octahedral holes in the lattice.

The transition metals of groups IIIB, IVB, and VB form metallic nitrides of the type Mn (M = Se, Ce, La Pr, Ti, Hf, V and Nb) and other nitrides (Mo_2N, W_2N, Mn_4N, Fe_4N). Interstitial or metallic nitrides are similar to interstitial carbides, borides and to interstitial hydrides though to a lesser extent. These nitrides have some relation to intermetallic compounds.

Some salient features of these nitrides are :

(i) They are extremely hard substances with high melting points. For example, tin nitride melts at 2947°C.

(ii) They have much in common with true metallic lustre, high electrical conductivity and other metallic properties.

(iii) Nitrides of composition MN possess the sodium chloride structure.

(iv) The properties of these nitrides are reflected in their crystal structures.

(v) These nitrides may be prepared by heating the metal to 1100 – 1200°C in a current of nitrogen or ammonia or they may be prepared by introducing a heated carrier filament of Pt, C or W into a mixture of nitrogen and the vapours of the metallic halide. The carrier filament is removed by volatilisation.

Being nitrogen deficient, these nitrides are often not exactly stoichiometric, and they are metallic in appearance, hard and electrical conductor, since the electronic band structure of metal persists. They are very stable and very inert, and do not hydrolyse except with difficulty. Again like carbides, they are extremely herd, and have very high melting points : *e.g.*, a representative compound VN melts at 2570°C, and has a hardness between 9 and 10.

For some elements, the nitride, oxide, and carbide are isomorphous, an example being TiN, TiO and TiC. The structure of nitrides is usually cubic like rock-salt. If the composition of the nitride is represented by Mn as ScN, LaN, CeN, TiN, VN, TaN, etc., the N atoms are in all the octahedral holes, but in the cases of Mo_2N atoms W_2N only one-half of the holes are occupied, and in the cases of Mn_4N and Fe_4N only one-quarter.